MW01621098
2120 W 69

TRUCKS
OF THE WORLD

TRUCKS OF THE WORLD

OVER 240 OF THE WORLD'S GREATEST TRUCKS

INGRID PHANEUF AND JAMES MENZIES

p

This is a Parragon Publishing book
First published in 2006

Parragon Publishing
Queen Street House
4 Queen Street
Bath BA1 1HE, UK

ISBN 1-40546-725-8

Editorial and design by
Amber Books Ltd
Bradley's Close
74–77 White Lion Street
London N1 9PF
www.amberbooks.co.uk

Project Editor: Sarah Uttridge
Design: Zoe Mellors
Picture Research: Terry Forshaw

Printed in Thailand

CONTENTS

WESTERN STAR

Introduction

Trucks are unlike cars—not just because they are larger and more powerful but because they exist for different reasons. While cars gained popularity soon after their introduction in the late nineteenth century as glamorous toys for the rich, trucks were of the masses and for the masses. They were ugly, noisy, and smelly and the only justification for their existence would be that they made sense. Economic sense.

Above: *The sloped hood of the Kenworth T2000 results in substantial fuel savings—an important consideration for long-distance haulers.*

Left: *The Western Star LowMax epitomizes what many North American owner/operators look for in a truck—a long, square hood, spacious sleeper cab, and plenty of chrome.*

Unfortunately, prior to World War I they didn't even really accomplish that. Before 1914, motorized commercial transportation in Europe, North America, and Asia was a rarity. Trucks were few and far between and generally used to attract the attention of potential customers rather than to increase earnings by increasing transportation efficiency. Motorized transportation would not truly catch on as an option for commerce, which had long since grown accustomed to horse-drawn wagons, trains, and ships, for quite some time.

Socially, trucks were considered annoying, if not downright dangerous, inventions. They were driven by socially-ostracized eccentrics and "mad" scientists, who saw a future where other, more sober-minded individuals saw none. Inventors and manufacturers were regulated to the point of near-extinction—banished to country roads where they would be neither seen nor heard. Actually using them to deliver goods to city dwellers was clearly out of the question. Still, the advocates of trucks would persist. They would attend national fairs and scientific gatherings in the hopes of winning prizes and/or the attentive ears of potential investors. But despite this, at the beginning of the twentieth century, very few business people or consumers had any use for trucks.

Trucks and war

The prejudice against trucks would change with the advent of large-scale industrialized war in Europe and Asia. Trucks would quickly prove their worth when large masses of supplies, arms, and soldiers had to be transported to and from battlefronts that were continually moving. Armed forces on all three continents found that such massive transportation needs could be much more quickly met by truck than by horse and wagon, giving them a competitive edge when it came to holding hard-won ground. And so the destiny of the world's very first steam truck—Cugnot's Steam Dray, constructed in 1796, would be fulfilled—sadly too late for the Frenchman, whose invention was ignored by Napoleon and who died in near-poverty, after a lifetime of exile from the country whose former king had supported him.

Despite Cugnot's best efforts, trucks would become creatures of the military in the popular imagination largely because of World War I, and the approbation of their countries' armed forces would give trucks the boost they needed in public opinion. Why, if they moved soldiers and their supplies so well, shouldn't trucks also be used to move consumer goods?

The commercial use of trucks would catch on slowly but steadily. With the construction of roads (by trucks) and the growing needs of consumers (fed by trucks), they would gain more acceptance. Horse-drawn carriages, even in urban areas, would begin to share the roads, at first grudgingly (some horse-driver unions actually encouraged their members to physically assault truck drivers), but with increasing equanimity. Meanwhile, trucks would also gain acceptance in rural areas where, thanks to attachments and fittings, they could facilitate work formerly performed by teams of horses and men. Other technological developments would further speed public acceptance. Steam engines would be replaced with electric and gasoline engines. Diesel would soon follow, making trucks less expensive to run. And payloads would increase, as truckmakers constantly strove to make lighter and more powerful vehicles.

In fact, trucks would do much to feed their own development, not only in terms of building roads, but also because they were proving to serve the needs of the consumer so efficiently (much more efficiently than rail or ship even) that the consumer would soon become completely dependent on them for all kinds of goods that had not even been regularly available previously. This dependency would increase as new technologies arrived to create and fulfill new needs. Refrigerated trailers brought frozen fish fresh from the sea to locations far inland, and oranges arrived in Canada in midwinter.

A second World War brought further developments and further needs, in terms of construction, to rebuild the destruction it wrought. Industry would grow to depend on trucks as much, if not more, than consumers. Trucks were needed not only for getting edible goods to market but also for transporting much-needed parts from one factory to another simply for assembly, never mind where the finished product would have to be transported afterwards.

Above: *Cugnot's development of the steam dray was funded by the French King Louis XV.*

Trucking today

We have a fully grown trucking industry today, inseparably intertwined with the fabric of our daily lives, so much so that when the cost of oil goes up, so does the cost of the goods we have become so accustomed to obtaining. In short, we depend heavily on truck transportation, and yet we have very little day-to-day awareness of it, beyond our annoyance and/or fear at being stuck behind a slow-moving cement mixer or a careering gravel hauler on the highway.

Perhaps our lack of appreciation for what trucks do for us has everything to do with how efficiently they accomplish the task they have been set. Still, if we weren't able to buy milk or fresh fruit at the supermarket, or fill up our cars at the gas station we would certainly notice it, but would we immediately associate its absence with a lack of transportation? Probably not. We'd probably assume there was a milk shortage, or that a factory was on strike somewhere, and we'd probably be right. Whether the transportation chain has broken is rarely in our thoughts, so we easily overlook the main purveyors of our goods, those big vehicles that get in our way on the journey to work in the morning.

"If you bought it, a truck brought it," is a saying that has been circulating through the industry for years, but one that rarely reaches the ears of outsiders. Given the fact that we are so blissfully unaware of the way trucks have become an integral part of our daily lives, is it any surprise that we are largely ignorant of the history of these vehicles, and how they are so interdependently linked with world's economic, industrial, and political development?

The history of trucks

Which came first, the chicken or the egg? This is the question that almost instantly arises when it comes to examining the role of trucks in the development of world economy, industry, and politics. Wars have been won and lost thanks to the strengths and failures of trucks. Populations have starved or died of exposure to the elements because trucks were either unavailable or unable to carry supplies to them. Trucks were also built to move entire populations, or to make sure entire populations didn't die of starvation and exposure. And infrastructures destroyed by war were completely rebuilt by trucks and jobs were provided by them. Economies have recovered and failed with and partly because of them. And most importantly, roads were built by them, roads that would eventually be traveled by them.

But trucks didn't just affect their surroundings; their surroundings also affected them. Commercial vehicles were a product of many factors beyond their control, and their great variance in terms of size, weight, strength, and function was the product of these external factors. Geography and climate played an important role. Trucks that developed in urban England were vastly different from those developed for mountainous Switzerland, icy Finland, or the rugged wet regions of the American Pacific Northwest. Crossing a desert required a different kind of vehicle than crossing a jungle did. Yet both journeys were accomplished by trucks, sometimes even trucks made by the same company, but with different engines, tires, and suspensions.

Below: *While most drivers were forced to sit out in the open air, the Volvo series 1 trucks could be fitted with a cab on demand. This was probably due to the cold climate in Sweden.*

Distances were different, and so were the road systems available when trucks made their maiden voyages. Social standards were different too—how difficult is it to imagine today that New Yorkers, now famed for their ability to live with almost constant noise and bustle of a city that supposedly never sleeps, objected so strongly to the noise and odor made by one truck inventor that the mayor actually banned him and his contraption from driving within city boundaries?

Insurance premiums had an impact as well—when companies started to refuse to insure steam trucks in North America and Europe due the fact that their boilers had a tendency to explode, truckmakers started to make the switch to other fossil fuels. Speaking of fossil fuels, availability of resources, including fuel and other raw materials, had an enormous impact on truck building, especially at the beginning of the twentieth century. The trucks of yesteryear were often built with whatever parts could be found, especially in times of scarcity, such as after the two world wars, when steel and gasoline were in very short supply.

Still, trucks were needed, and truckmakers found ways and means of providing them. It is a testimony to their ingenuity that they did. And it was essential to their survival as well: If they hadn't been able to change with the times and adjust to the needs of industry and consumers, many of them wouldn't exist today.

Political influences

It might surprise some to learn that politicians have always had a great, though covert, interest in trucks and transportation issues. While much of society continues to turn its nose up at trucks—they are still considered smelly, noisy, and dangerous—politicians understand their importance and recognize them as one of the forces that power the world's economy.

Politicians battle citizens to widen roads through established neighborhoods to accommodate trucks. They make laws that enable trucks to move down highways and over bridges more quickly and efficiently, and try to forge international agreements that make it easier for trucks to cross borders.

The future of trucks

Trucks still vary in their design according to the availability of raw materials. Due to growing difficulty of obtaining fossil fuels, the cost of operating trucks is growing by

Below: *Trucks like this GMC CCKW 6x6 served a vital role in wartime, transporting troops and munitions.*

Above: *Cabover engines such as the Volvo FH16 continue to be the models of choice in Europe. Volvo's flagship truck took the horsepower wars to a new level with up to 610 hp (454.8kw) available and the industry's highest torque.*

the day. Truckmakers know this and build their trucks accordingly, with auxiliary power units and heaters in the cab that provide heat and/or lighting to the cab when the engine is turned off to save on fuel consumption, and with lighter parts that also reduce fuel consumpti on and increase payload. Furthermore, with increasing worldwide concern about global warming and air pollution, engine- and truckmakers are now racing to meet increasingly demanding requirements for the reduction of engine emissions. The result is that in the future, trucks will be emitting exhaust fumes that are cleaner than the air they take in.

Trucks today are still at the forefront of technology. Communications in trucks and between onboard computers and drivers and their distant headquarters are equally evolved, to the point where stolen trucks can actually be tracked by satellite and drivers can be identified by their fingerprints. And other technological innovations now allow for trucks to actually correct themselves when they are on the verge of overturning while cornering, or keep an electronic eye on whether their driver is getting sleepy.

Driverless trucks?

Trucks are still dependent on drivers to make them run efficiently. (No doubt there is a truck technician somewhere developing the technology to replace drivers, too.) But drivers and owner/operators are perhaps the only human beings left who truly understand and appreciate the truck and its history. As for the uninitiated, welcome to the world of trucks!

DOWNASH & PEVENS
TRACTION Co
HAILSHAM SUSSEX
U.W. 5 TONS 13 CWT 2 Q
A.W.F. 4 TONS 5 CWT 0 QTRS
A.W.R. 8 TONS 10 CWT 0 QTRS

chapter 1

The First Trucks

The first trucks were neither very efficient, nor were they very esthetically pleasing. Their development was driven by the vision of eccentrics who could see a future for them when no one else could or even wanted to, and who had the time and/or inclination to tinker. These individuals, more than anything else, drove truck development forward to the point where they became a necessity rather than a nuisance.

Above: *The era of the modern truck begins in 1885 with the introduction of the four-stroke internal-combustion engine.*

Left: *Fodens Ltd. were pioneers of steam, having built their first steam wagon in 1898. This overtype dates from around 1912.*

Several problems faced those who would try to convince the world that the truck had a future. First there was the problem of how to power the engine. Steam power was dangerous to the driver and the vehicle's cargo. Electricity was problematic, because of the need for frequent recharging of batteries, which constituted a significant payload.

Technology aside, there were plenty of simple physical problems preventing the widespread use of trucks as well. Lack of roads was one thing, as well as the lack of maintenance of those roads that did exist. Truck tires just couldn't fit into the ruts left by horse-drawn carriage wheels, and they had trouble navigating the rugged surface of unpaved roads. Furthermore, with the distances traveled by most people being so short, the need for trucks traveling at greater speeds than horses was not that obvious. Replacement parts and maintenance were also an issue. Early trucks were handcrafted and unique. New parts would not only have to be ordered, they would have to be built. And there was the problem of fuel—where to get enough coal, wood, electricity, gasoline, or diesel to power the engine.

In short, there were many elements, social, physical, legislative, and even political that would have to fall into place before the use of trucks could truly become advantageous.

CUGNOT
1769 **France**

CUGNOT'S STEAM DRAY

Built by French army officer Nicholas Cugnot, the "grandfather" of the modern truck, this steam-powered truck was intended to transport cannons for the French army.

Inset: *Nicholas Cugnot's steam dray could carry up to 5 tons at 2 mph (3.3 kph) but could only generate enough steam to travel about 700 yards (640m) before stopping to build up more steam.*

Above: *With a heavy boiler mounted ahead of its single front wheel Cugnot's vehicle was inherently unstable and is said to have collided with a stone wall while being demonstrated. Consequently its driver was responsible for the first recorded traffic accident.*

The steam dray had three wheels with iron rims, two at the back and one in front. The steam power came from a boiler hooked up to a two-cylinder engine, both of which were mounted over the front wheel of the vehicle. It could travel at 2.48mph (4km/h), stopping every 10 to 15 minutes to build up enough steam pressure to get going again.

In 1771, Cugnot built a second vehicle for testing which was a tractor designed for a 4–5 ton payload. Cugnot sought the approval of France's Minister of War at the time, the Marquis de Monteynard, for some secret tests to be staged at Versailles. Secrecy was necessary at this time because of the fact that Cugnot and some of his highly placed supporters at the court of Louis XV viewed the steam dray as something of a national military secret. Despite this, the tests were never conducted, because the Minister did not give his approval or even answer Cugnot's letters.

First truck accident

Tests were conducted, nevertheless, and the vehicle was soon credited with causing the world's first traffic accident by colliding with a garden wall. In spite of such accidents, one of Cugnot's vehicles survived the French Revolution and today resides in the National Academy of Arts and Sciences in France.

Napoleon took little interest in steam-powered gun tractors, so Cugnot and his steam-powered vehicles sadly faded into obscurity immediately after the Revolution.

Specifications

Country: France
Year manufactured from: 1769
Engine: two cylinder, steam powered
Transmission: not available
Payload: 4 tons (4 tonnes)
Applications: cannonball carrier
Special features: one powered wheel in front

TREVITHICK
1803 UK

LONDON STEAM CARRIAGE

In 1803, Richard Trevithick and his associate Andrew Vivian designed the London Steam Carriage, a coachlike, steam-powered vehicle capable of going up to 9.9mph (16km/h).

Right: *Though built to carry passengers, Trevithick's London Steam Carriage could, in theory, have been adapted to carry up to half a ton of goods. However, it would have been rather top-heavy and, being a tricycle, not too stable.*

James Watt was the first to patent "parallel motion" in steam engines, which could keep a rigid piston rod moving vertically while attached to the end of an oscillating beam. He was also one of the first to use a definite measurement to compare steam-power to horsepower. One horsepower was exactly 33,000 foot-pounds of work per minute. Within a few years, Watt's engines were frequently referred to as 14-horse engines, 20-horse engines, and so on.

The London Carriage

Following on from Watt's ideas, in 1803, Trevithick (who also worked with Watt) and his associate Andrew Vivian assembled the London Carriage at Felton's carriageworks in Leather Lane. On completion, the carriage was driven about 10 miles (16km) through the streets of London with seven or eight passengers, in an attempt to demonstrate to the public that self-powered passenger and/or cargo vehicles were a possibility.

While the experiment was considered successful, the vehicle proved highly unpopular, mainly because it terrified horses, but also because of several deaths due to horrific boiler accidents. There was also the issue of fires caused by sparks from the funnels. The media and public largely ignored the London Carriage and the vehicle was eventually dismantled.

Specifications

Country: UK
Year manufactured from: 1803
Engine: two cylinder, steam powered
Transmission: not available
Payload: not available
Applications: experimental
Special features: first self-powered passenger carrying vehicle in the world

DUDGEON
1866 USA

DUDGEON STEAM WAGON "RED DEVIL"

In 1855, Richard Dudgeon astounded New Yorkers by driving from his home to his place of business in a steam carriage. The noise made by the truck was so bad that city authorities confined it to one street.

Specifications

Country: USA
Year manufactured from: 1866
Engine: steam powered
Transmission: not available
Payload: not available
Applications: experimental
Special features: driver at the back

With cedar wheels and iron tires, the "Red Devil Steamer" met with much criticism. The wagon was apparently just too noisy and smelly for the delicate sensibilities of New Yorkers at that time.

Dudgeon had constructed the "Red Devil" in 1866, after losing his first steam wagon in a fire. Public opposition to the vehicle drove Dudgeon, his family, and his creation to Long Island, to escape city officials. Here he and his carriage became a familiar sight.

Dudgeon claimed to have run the steam carriage many hundreds of miles. He also claimed to have once covered a mile in less than two minutes. He reported the vehicle could carry as many as 10 people and run for four hours at 14mph (22.5km/h) on just one barrel of anthracite coal. The vehicle reportedly weighed 3,700lb (1,678kg) with enough water and fuel to run for one hour. Even so, it failed to gain public favor.

Public not interested

Richard Dudgeon wrote in 1870 concerning his Steamer, "After seventeen years of effort and conviction of its utility, I have learned that it is not fashionable, or that people are not ready for it." In spite of all his claims, the disillusioned inventor of the steam carriage had eventually became reconciled to the fact that the public was just not interested.

Above: *After losing the original in a fire, Dudgeon constructed a second steamer in 1866. But there was much opposition to the vehicle being driven on the streets of New York.*

BOLLÉE

1873 France

L'OBÉISSANTE

Amédée Bollée was one of the most ingenious French engineers of his time. Despite its old-style design, *L'Obéissante* was a very advanced steam vehicle.

Right: *Frenchman Amédée Bollée did much to advance steam-carriage design with his* l'Obéissante, *which was the first vehicle to feature Ackerman-type steering controlled by a steering wheel.*

Specifications

Country: France
Year manufactured from: 1873
Engine: steam boiler with two cylinders
Transmission: not available
Payload: up to 12 passengers
Applications: utility steel truck
Special features: first steering wheel, independent front suspension, full forward control

Amédée Bollée was born in 1844. His father was a bell founder in the city of Le Mans. At first, the young Amédée worked in the foundry, where he perfected a new casting process. In 1867, he visited the French Exposition Universelle in Paris. There he saw steam carriages, and as an engineer took interest in this vehicle and began to work on a new project. Three years after the 1870 Franco-Prussian war, the steam carriage *l'Obéissante* (the Obedient One) was ready. The vehicle contained real craftsmanship, with accommodation for 12 passengers, a roof, and windows, the overall style being like that of a train passenger car. The quiet steam boiler at the rear was another masterpiece. It burned 110lb (50kg) of coal per hour and the two cylinders under the floor could move *l'Obéissante* from 9 to 25mph (15 to 40km/h). This amazing top speed (for the time) was offset by a small range of 16 miles (25km). But the most advanced design within the steam carriage was the steering wheel, the first one in the automotive industry.

Disappointment

After trials, *l'Obéissante* proved itself a very capable vehicle, and at the end of 1875 it made Le Mans to Paris (114 miles/183km) in only 18 hours. Bollée, however, was unable to sell the vehicle.

BOLLÉE

🛠 **1879 France**

MARIE-ANNE

After *L'Obéissante*, Amédée Bollée worked on a lighter steam car called *La Mancelle*. In 1879, however, he returned to a monster steam road train.

Specifications

Country: France

Year manufactured from: 1879

Engine: steam boiler with two cylinders

Transmission: three-speed

Payload: up to 100 tons (98 tons) towed

Applications: special big hauler

Special features: six-wheeler steam hauler with articulated four-wheel drive

Above: *This strange steam monster was manufactured for a French foundry company that needed a road train to carry its materials. Note the steam boiler at the rear and the two-wheel drive trailer.*

The *Marie-Anne* was ordered by the Société Métallurgique de l'Ariège, a foundry company. This steam road train weighed 19.6 tons (20 tonnes) and could haul up to 98 tons (100 tonnes). There were six wheels, including four wheels providing drive. The steam boiler was at the rear and the front cylinders drove the wheels through a complex system of shaft drives and external chains. The two-wheel drive trailer held the water and coal supplies. A three-speed variator mechanism was designed to maintain constant speed.

Interior Comfort

There were passenger seats at the front, near the steering wheel, but the stoker had to stand up. During transit the *Marie-Anne* was outstandingly quiet, except when the steam road train had to stop to "refuel," taking on water in the villages. Wherever it stopped, the steel monster was a star attraction. The public and the authorities were very impressed by the machine, and the *Marie-Anne* was favorably considered for production. At this time, Bollée had signed some important contracts in Germany and a sister ship of the *Marie-Anne*, called *l'Elisabeth*, was sent into this country. But Bollée's financial situation was so disastrous that he finally stopped the assembly lines. The final vehicle was a mail coach ordered by the Marquis de Broc. Amédée Bollée passed away in 1917.

BENZ

1885–89 Germany

BENZ MOTOR WAGEN

Unlike steam engines, the internal-combustion engine operates by burning fuel inside the engine. With the introduction of the four-stroke internal-combustion engine from 1885, the era of the true truck begins.

Right: *Carl Benz perfected the first practical internal combustion-engined vehicle—a three-wheeled "motor wagon." It was powered by a horizontal single-cylindered engine which could run on gasoline, paraffin, or naptha.*

Specifications

Country: Germany
Year manufactured from: 1885
Engine: four stroke internal combustion powered by paraffin, gasoline, or naptha
Transmission: not available
Payload: not available
Applications: experimental
Special features: three wheels

The forerunner of trucks powered by internal combustion engines was a Benz three-wheeler. It could run on paraffin, gasoline, or naphtha, and could reach 9.3mph (15km/h). Benz's first motorized vehicle eventually featured three innovations still in use in trucks and cars today—an electric battery, spark plug, and electric coil ignition—but not from the very beginning. Benz's first electric ignition used a generator, but the generators of the early 1880s were not up to the task, so the inventor shifted to a storage battery. The vehicle's spark plug consisted of two pieces of insulated platinum wire stuck into the combustion chamber.

No fuel injection

The fuel induction system also left something to be desired. The engine didn't use fuel injection or anything resembling a true carburettor. Instead, fuel drained into a can filled with fabric fibers and then evaporation carried the fumes into the cylinder.

Even so, the vehicle's four-stroke engine worked, the four strokes of the engine consisting of intake, compression, power, and exhaust. And after a major reorganization of the company in 1890, the Benz motor works was on its way to becoming a successful manufacturer of vehicles.

PEUGEOT
1895 France

PEUGEOT TYPE 13

The Peugeot brothers introduced five units of the Type 13, the company's first gasoline-powered truck, in 1895. The truck was a light delivery vehicle, with a Daimler internal-combustion engine.

Specifications

Country: France
Year manufactured from: 1895
Engine: gas powered internal combustion
Transmission: chain drive
Payload: 882lb (400kg)
Applications: light delivery truck
Special features: first commercial vehicle by Peugeot

Above: *France's earliest light delivery truck was based on the Peugeot Type 13 of which five were produced as early as 1895. They were powered by Daimler gasoline engines.*

In 1888, Armand Peugeot's fascination with motorized vehicles was born. He very proudly unveiled a Peugeot steam-powered tricycle at the Paris World Fair. But he also started to experiment with gasoline-powered engines, and produced four units of the first gasoline-powered four-wheel car, called the Type 2, fitted with a Daimler engine at the Valentigney factory.

It would soon include solid rubber or optional Michelin tires with rubber tubes, thanks to Edouard Michelin, who tried out his pneumatic tires on the company's Éclair model car in the Paris–Bordeaux–Paris race of 1895.

Heavy-duty production

By 1900 Peugeot was manufacturing several heavy-duty vehicles (for that time) including the Type 18 eight-seater bus with proprietary Peugeot engine (Armand stopped buying Daimler's engines from Panhard and Levassor in 1896), the Type 22 two-seater pickup truck, and the Type 32 wagonette-tonneau (21 units in 1900). The Type 34 and 35 commercial vehicles quickly followed. The Type 36 single-cylinder was the first tonneau with a hood at the front and a steering wheel with an inclined steering column instead of handlebars. It marked Peugeot's move to chainless transmission.

LEYLAND STEAM VAN

1896 UK

LEYLAND STEAM VAN

Above: *Leyland's early steamers were oil-fired as opposed to coal-fired. This is their first wagon, built in 1896.*

Leyland has a long and distinguished history dating back to 1896, when the Sumner and Spurrier families founded the Lancashire Steam Motor Company in the town of Leyland, in northwestern England.

Right: *Leyland continued to produce small numbers of steamers until the mid-1920s, but the company focused more on the internal combustion engine starting in 1904.*

Specifications

Country: England
Year manufactured from: 1896
Engine: two cyliner steam
Transmission: three-speed
Payload: 1.5 tons (1.016 tonnes)
Applications: delivery
Special features: forward control, underfloor engine.

The company's original product was a steam delivery van with a capacity of 1.5-ton (1.016-tonne). The company also soon came out with a 2-ton (2.03-tonne) version, which proved popular. Steam wagons for carrying freight were very successful in the early 20th century, thanks to their low running costs, which enabled them to survive the advent of gas-driven trucks.

Nevertheless, Leyland had the gift of prescience and started producing gasoline-powered vehicles as well as steamers early on.

In 1907 the company was manufacturing 35 steam-powered machines and 17 that ran on gasoline. The final and perhaps fatal blow to steam-powered trucks in Britain came in the 1920s when the government introduced a new regime of road taxes based on the weight of a vehicle when unloaded. This put the water-bearing steam wagons at a further disadvantage compared to the lighter gasoline trucks.

DAIMLER
1896 Germany

THE DAIMLER MOTOR LASTWAGEN

The first gasoline-powered Daimler Motor Lastwagen was completed on October 1, 1896. The truck was entered in the production records of Daimler-Motoren-Gesellschaft in Cannstatt, near Stuttgart, Germany.

This truck looked like a horse-drawn cart without a drawbar (the metal or wooden bar attached to the front of the carriage to which the horse was attached). The driver sat in front of the front axle, on a coach box, and in the open air, without any protection from the elements. The engine was situated at the rear.

The truck which was created by Gottlieb Daimler and his designer and protege Wilhelm Maybach, was powered by a "Phoenix" two-cylinder in-line engine (also known as model "N") which could push the vehicle at up to 8mph (12.8km/h). Engine power reached the rear axle via a belt system.

Engine cooling system

The Daimler Motor Lastwagen also included a rotational engine cooling system. Water was admitted to the engine's interior flywheel and rotated, then returned by centrifugal force to the engine's water jacket via a pipe. There was also a heater that could be mounted on the driver's box, which also ran on circulating cooling water.

Later innovations put the engine under the floor and beneath the driver's seat, allowing for more loading space. The driving components of the belt drive were also placed under the loading space, and the drive eventually consisted of pinions engaging internal-toothed gear wheels attached to the

Above: *While the engines of the first Lastwagens were placed behind the driver, later innovations to the Lastwagen would put the engine under the floor and beneath the driver's seat, allowing for more loading space. The man on the right is Gottlieb Daimler.*

two rear wheels. Daimler and Maybach's truck soon became available in 4, 6, 8, and 10hp (2.9, 4.4, 5.9, and 7.4kw) versions, with payload capacities between 3,300 and 11,000lb (1,500 and 5,000kg) and top speeds of up to 7.5mph (12km/h).

The DMG eventually opened a truck plant in Berlin, because Gottlieb Daimler's partners were based there. They built automobiles and trucks under the Daimler licenses, many with electric motors. Gottlieb Daimler died in 1900, but his trucks lived on when, two years later, the companies in Cannstatt and Berlin were merged under the DMG (Daimler-Motoren-Gesellschaft, Daimler Motor Company) banner. The Berlin factory developed into the commercial vehicle plant of DMG. Models in the early 1900s included a new 5-ton (5-tonne) chassis made of sheet steel with riveted frame members and a three-tonner with a 28hp (20.8kw) engine, an enclosed gearbox and automatic central lubrication.

Trucks for the German army

German trucks gained ground while still in the early stages of their development thanks to the army, which put in orders for military motor vehicles, complete with specifications for size, design, and equipment, as early as 1898. When World War I broke out, there were 5,000 subsidized trucks at the German army's disposal, including the five-tonners, with 22 and 35hp (16.4 and 26kw) engines from DMG in Berlin and 28 and 40hp (20.8 and 29.8kw) from the Benz-Werke in Gaggenau.

Incidentally, Carl Benz, for whom the Gaggenau production facility was named, was developing vehicles similar to Daimler's at roughly the same time. The two never met but they were probably acutely aware of each other's work. Benz's three-wheeled Patent Motor Car and Daimler's motorized carriage were produced about 10 years prior to their introduction of trucks to the market, and Benz's first "combination delivery vehicle" (a van) and Daimler's first truck both came out in 1898 and 1896 respectively.

First combination vehicle

Benz had fitted a box body on the frame of one of his cars. The payload of the fully loaded four-wheeled vehicle was 662lb (300kg) plus the driver. Its single-cylinder engine could provide 2.75hp (2kw). A year later, the "delivery vehicle" could carry the same load plus two people thanks to its new 5hp (3.7kw) single-cylinder engine. In 1900 Benz would launch an entire heavy-duty truck series, with a gross weight of up to 5 tons (5 tonnes).

Inset: *The Daimler Motor Lastwagen also included a rotational engine cooling system. Water was admitted to the engine's interior flywheel and rotated, then returned by centrifugal force to the engine's water jacket via a pipe.*

Specifications

Country: Germany
Year manufactured from: 1896
Engine: gas powered two cylinder in line
Transmission: belt driven
Payload: 5,940lb (2,700kg) fully loaded
Applications: delivery
Special features: underfloor rear engine, forward control.

Below: *A "Phoenix" two-cylinder in line engine (also known as the model "N") powered the Lastwagen and could push the vehicle up to 8mph (12.8km/h).*

W.J. STILL
1899 Canada

W.J. STILL ELECTRIC DELIVERY WAGON

Less than a year after Robert Simpson purchased his Fischer No. 2 Coach Delivery Wagon, Parker Dye Works purchased a Toronto-made electric delivery wagon from the W. J. Still Motor Company.

Specifications

Country: Canada
Year manufactured from: 1899
Engine: battery powered electric
Transmission: belt driven, rear-wheel drive
Payload: not available
Applications: delivery
Special features: Ackerman-type steering

The wagon surely provoked the envy of the driver of the Fischer wagon, because the driver of the Still wagon was provided with lower seating under a protective roof. The motor was midmounted, with a belt driving the rear wheels.

Ackerman steering

The Still wagon was also more stable, thanks to the inclusion of Ackerman-type steering. The Ackerman Steering Principle defined the geometry applied to all vehicles (two- or four-wheel drive) to enable the correct turning angle of the steering wheels to be generated when negotiating a corner or a curve. Before the principle was developed, horse-drawn carriages had parallel steering arms and suffered from poor steering performance (tending especially to tip while cornering at high speeds). Rudolf Ackerman is credited with working out that using angled steering arms would cure these vehicles of such steering problems.

Electric cars and trucks enjoyed popularity both in North America and Europe at the turn of the century.

Above: *Like all electric vehicles the Still was limited to local deliveries because its batteries needed regular recharging. It did, however, boast some advanced features including Ackerman-type steering.*

WHITE STEAM
1901 USA

WHITE STEAM DELIVERY VAN

Above: *The White family's love affair with transportation began when the company started manufacturing bicycles. Rollin White soon developed a steam engine.*

White steam trucks were already popular and used for a variety of applications in the early 1900s. Their light-duty chassis could be fitted for various tasks—from fire trucks to delivery vans and funeral hearses.

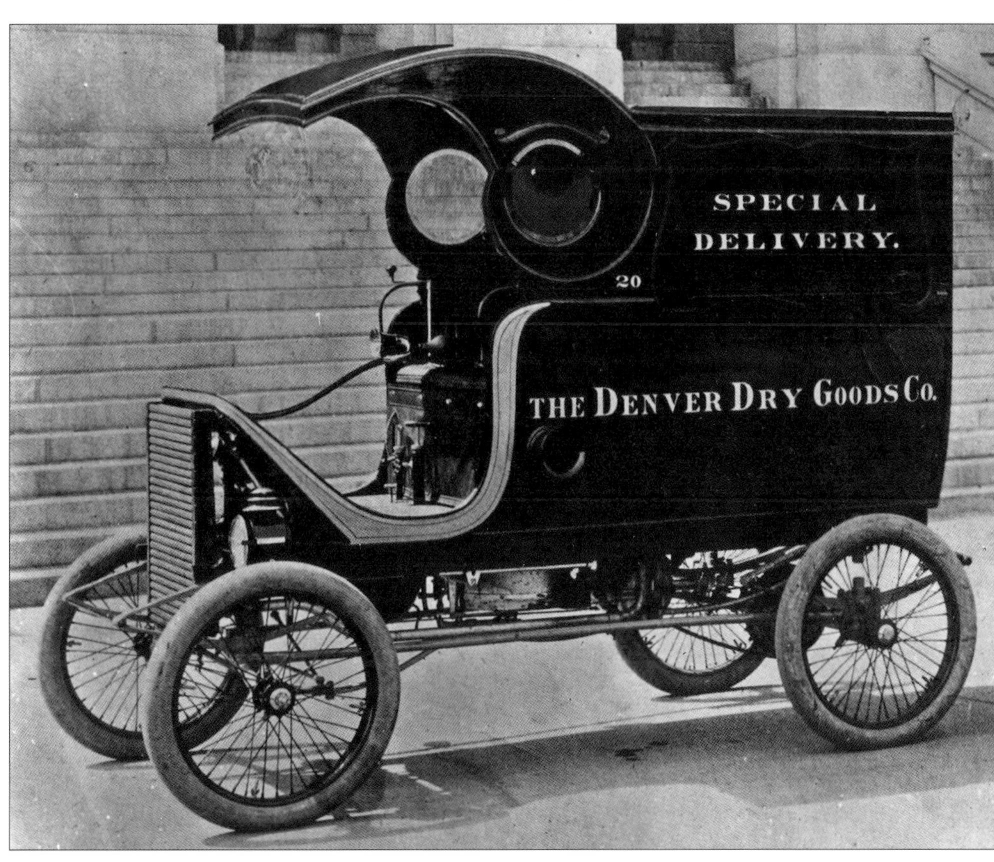

Right: *By the early 1900s, the brothers had brought out a car, as well as a small delivery truck.*

Specifications

Country: USA
Year manufactured from: 1901
Engine: steam powered
Transmission: belt driven
Payload: 441lb (200kg)
Applications: light delivery
Special features: flash boiler was safer and faster than existing boilers

The company was founded by Thomas White in Massachusetts, first making sewing machines. When automobiles began making an appearance, Thomas White's sons, Windsor and Rollin, convinced their father to let them start developing one of their own. The two brothers saw a whole new market opening up, and set up an Automobile Department at White. White produced its own automotive components, and Rollin developed a proprietary White steam engine. By the early 1900s, the brothers had brought out a car, as well as a small delivery truck. The White brothers soon split off from the Automobile Department to form their own White Motor Car Company.

Gasoline replaces steam

Soon gasoline replaced steam power, and White quickly adapted, continuing to build cars and light trucks. During World War I, the company produced trucks for the military at home and abroad. White trucks proved to be as much of a success overseas as they had in North America.

FODEN
1899 UK

FODEN STEAM WAGON

E. Foden and Sons was a family-run business with a history dating back to the steam engines of the 19th century. The company built its first steam wagon or truck in Sandbach, in Cheshire, England in 1899.

Above: *Fodens Ltd. favored the "overtype" steam wagon with locomotive boiler. The two-cylinder engine sat above the boiler as on this 5-tonner, which dates from before World War I.*

Developed from the beginning of the 20th century, the first steam wagons and trucks were called "overtypes," with their engines mounted on top of the boiler. The engines were chain driven and capable of speeds up to 30mph (48km/h). Designs included four- and six-wheelers, artics, and tippers.

Prior to 1899, Foden had already acquired a fine reputation for its steam tractors (built for agricultural work), but wanted to start producing wagons as well. An opportunity to do so arose when the War Office ran a trial for steam lorries needed in South Africa for the Boer War.

There were several contenders. Foden's entry, wagon works number 524 (504 having been a prototype) completed the 257-mile (414km) course in 41 hours, 41 minutes, compared with the next fastest wagon at 44 hours, 46 minutes. The Foden 524 used 3,111lb (1,410kg) of fuel and 1,918 gallons

(8,719 liters) of water, compared with the next best of 5,079lbs (2,303kg) and 3,060 gallons (13,910 liters), making it the clear winner according to some present. Foden's entry nevertheless came in second, due to what was judged the superior cross-country performance of the Thornycroft steam wagon. Both wagons were shipped to South Africa, however, where they were operated by 45 Company Royal Engineers. Thus Foden's first steam wagon was born.

First wagon sold in 1902

Silver Springs bought Foden's first commercial steam wagon, the M252, with a timber body by Jennings of Sandbach, in early 1902. The cloth-dyeing company wanted a reliable form of transport to haul products back and forth from its Cheshire plant to Manchester, and horse-drawn wagons were just too slow. The M252's delivery to the Silver Springs factory was occasion for much fanfare, at least according to the local press, which reported the wagon was followed by a procession of supporters all the way from the Foden factory in Sandbach to Timbersbrook, and greeted on its arrival by a brass band. A dinner party ensued.

Later in 1902, the M251 followed, and the two trucks did the Manchester run for Silver Springs on alternate days, leaving at 4 a.m., and returning at about 8 p.m. Another Foden steam wagon did the local coal run to Brunswick Wharf. The M1737 (faster and with rubber tires) joined the fleet in 1907. Records show that M252 was still going strong until 1926, when the factory exchanged it for a newer diesel-powered model from Foden.

Other steam-powered Fodens

But there had been other steam-powered Fodens in the meantime, including a 1914 short-chassis model, which would be fitted out with an open-backed tipper wagon application.

The "undertype" wagon was another variation—it was made with a vertical boiler with the engine mounted under the chassis, similar to a modern truck. Later models were fitted with pneumatic tires and could reach speeds of up to 60mph (96km/h).

Specifications

Country: UK
Year manufactured from: 1899
Engine: steam powered
Transmission: three-speed
Payload: 4 tons (4 tonnes)
Applications: delivery, military
Special features: engine positioned over boiler

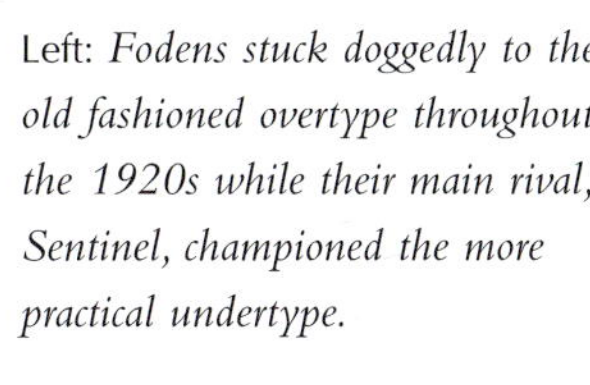

Left: *Fodens stuck doggedly to the old fashioned overtype throughout the 1920s while their main rival, Sentinel, championed the more practical undertype.*

SAURER
26

chapter 2

Commercial Vehicles

The opening years of the 20th century were just slightly more auspicious for truckmakers than the end of the 19th century had been. There was a general move to the production of gasoline-driven commercial vehicles.

Above: *SAF made commercial vehicles in Gaggenau, Germany. The trucks were said to be ahead of their time because of their innovations.*

Left: *Saurer trucks were so popular during this period that by 1914 they had sold 130 different models in Europe.*

Mass production of trucks and other vehicles was becoming established, making repair and maintenance that much easier for owners. International travel led to international trade in the trucking business, which meant a manufacturer could find a buyer a continent away. Truckmakers hitherto isolated in their own countries started to make alliances to sell and/or assemble each other's vehicles. For example, as early as 1909, FIAT brand cars and trucks were being built in New York. Diesel, which made running a truck even cheaper, was also gaining ground in Europe.

Furthermore, truckmakers had finally caught onto the fact that their trucks would have to do more than just travel from point A to point B with a payload to be considered useful, so trucks were customized. Truckmakers started attaching tippers and cranes. Some were lighter, heavier, faster, and stronger—they could come either with bodies or without, or with or without cabs. They could be used for hauling, digging, building, dumping, putting out fires, as ambulances, or to transport large human or animal cargoes. Trucks could do anything and manufacturers competed fiercely to prove they could tailor their product to almost any use. Companies strove to demonstrate their product was the best in its class.

RAPID

🛠 **1901 USA**

RAPID TRUCK

In 1902, Morris Grabowsky, with his brother Max, started one of the first truckmaking companies in the United States—the Grabowsky Motor Vehicle Co. of Detroit, Michigan.

The Grabowskys had already designed and built their first truck, also called the Rapid, in 1901. The very first gasoline-driven truck to run on the streets of Detroit, the 1901 Rapid was a light delivery van equipped with a one-cylinder engine and chain drive. It rode on a wheelbase of 80in (2.03m) and weighed about 1,900lb (861kg). The first Rapid was sold to the American Garment Cleaning Company for $1,250.

By 1905, the brothers had already built and sold about 75 trucks and several automobiles. In 1904, they changed the firm's name to Rapid Motor Vehicle Co. and in 1906, they moved their successful company to a new plant in Pontiac, Michigan.

Above: *Max and Morris Grabowsky completed their first truck called the "Rapid" in 1901 to become one of North America's earliest truck manufacturers.*

By then Rapid Truck company was specializing exclusively in the design and manufacture of commercial vehicles (one of North America's first truckmakers to do so) and a fact that was omnipresent in the company's marketing strategy of the time.

A print advertisement for one Rapid truck of the early 1900s (1906 to be exact) read as follows: "Our entire organization has from the very first been devoted exclusively to the building of Commercial Vehicles.... The success of 'Rapid' motor wagons is due to the fact they are built with the idea of giving heavy, continuous services, and carrying heavy loads."

By 1906 the Rapid Motor Company's truck chassis were available in a range of sizes, from 1 to 1.5 tons (1.016 to 1.52 tonnes), and with the buyer's choice of bodies. Trucks like the 1906 wagon weighed as much as 2,230lb (1,011kg) and could carry up to 3,000lb (1,360kg). As was the case for trucks designed and manufactured at that time, the rear wheels were larger than the front wheels and the tires were solid rubber. The 1906 trucks had also been beefed up to include double-sided chain drives to insure maximum power delivery to the 2in (5cm) solid steel axles. They went from a half-ton truck to a one-and-a-half tonner.

Below: *Within just four years no fewer than 75 Rapid trucks had been produced.*

Specifications

Country: USA
Year manufactured from: 1901
Engine: gasoline powered, one cylinder
Transmission: chain drive
Payload: up to 3,000lbs (1,360kg)
Applications: light delivery
Special features: forward control, underfloor engine

FIAT

1902 Italy

FIAT DELIVERY VAN

FIAT had its start in July 1899 when the *Societa aninima Fabbrica Italiana di Automobili—Torino* (FIAT) was formed. FIAT's first factory was located on Corso Dante in Turin, Italy.

Above: *By 1910 Fiat was making hooded-engine trucks like this conventional truck.*

The company's first vehicle (about 25 of them were produced) was a 3.5hp (2.6kw) gasoline-driven car, capable of reaching speeds of up to 21.7mph (35km/h). Other models soon followed, in 8, 10, and 12hp (5.9, 7.4, and 8.8kw) versions. Among them was the first FIAT with its engine in front (over the front axle). The front-engined FIAT had a two-cylinder 64.8ci (1,082cc) engine producing 10hp (7.4kw). The vehicle was so popular, the company built one per week (this was prior to mass production). The 12hp (8.9kw) model, for its part, was the first FIAT to be widely exported, selling throughout Europe and to the United States. A total of 134 were built.

By the early 1900s, the company had diversified into the production of trucks, buses, trams, taxicabs, and marine and aero engines. After going public in 1902, FIAT had established a number of new companies serving specific functions, including the Societa Carrozzeria Industriale for commercial chassis production.

First petrol-driven truck

The company's first commercial vehicle, a gasoline-powered delivery van, was built in 1902. The van was powered by an 8hp (5.9kw), 115.3ci (1.9-liter) vertical twin-cylinder engine, with a three-speed gearbox and a side-mounted chain drive.

A heavier version came soon after, in 1903, with the introduction of a 4-ton (4.06-tonne) truck that rolled (one would suspect somewhat noisily) on iron tires.

Specifications

Country: Italy
Year manufactured from: 1902
Engine: gasoline powered, twin cylinder
Transmission: three-speed, chain drive
Payload: not available
Applications: delivery
Special features: chain drive was side mounted

SCANIA

1903 Sweden

SCANIA TRUCK

Scania started out making bicycles and motorcycles, but had begun manufacturing cars by 1901. Two years later, the company produced its very first in a long line of trucks.

Above: *Scania first entered the commercial vehicle scene with this chain-drive 1.5-tonner powered by a 12 hp (9kw) twin-cylinder gasoline engine.*

Specifications

Country: Sweden
Year manufactured from: 1903
Engine: two cylinder Kamper
Transmission: chain drive
Payload: 1.7 tons (1.72 tonnes)
Applications: delivery
Special features: engine mounted under driver's seat

Scania's first commercial vehicle had a two-cylinder 10hp (7.4kw) Kamper engine mounted under the driver's seat, with a chain drive turning the wheels. It also had a 1.7-ton (1.72-tonne) payload. One can only speculate as to why, but Scania obviously felt it could improve on the Kamper engine. By the end of 1903, the company had begun designing and manufacturing engines of its own. Now, all the company had to do was create a market for its trucks.

Creating a market

By 1907, Scania (and Vabis, another local truck manufacturer with which Scania would later merge) had developed an engine-driven tipper platform. Such developments emphasized the functionality of Scania trucks—and how they could ease the workload of laborers, thus maximizing manpower. Yet the company needed to capture the imagination of the general public. So a Scania truck was driven from Malmo to Stockholm, a journey calculated to showcase the vehicle's abilities. At 12.4mph (20km/h), the 435-mile (700km) journey took only three days. But the two-axle truck would consume more than 87 gallons (400 liters) of fuel. One can't help but wonder if the overall fuel consumption of the vehicle was kept quiet by the company.

SAURER
1903 Switzerland

SAURER FIVE-TONNER

The first Saurer truck was built by Adolph Saurer at his factory in Arbon, Switzerland, in 1903. It grossed 5 tons (5.08 tonnes), and was powered by a four-cylinder, T-head, 25 to 30hp (18.6 to 22.3kw) engine.

Above: *The 1903 Saurer 5-tonner was unusual for a heavy truck of that period in featuring shaft drive. Soon after production commenced, however, this was changed to chain drive.*

The truck proved popular, collecting trophies at several international contests. Factories were erected in Paris and Lindau, Germany, and the company started distributing in Russia, Canada, and Japan. The trucks first appeared in the U.S. in 1911. By then Saurer trucks ranged from 1.5 to 3.5 tons (1.52 to 3.55 tonnes), all of them with four-cylinder engines producing 16 to 30hp (11.9 to 22.3kw) and four-speed transmissions. The smaller trucks were shaft-driven, while the larger ones used chains.

The trucks were so successful in Europe that by 1914, Saurer was producing 130 different models. The company's success continued into World War I, when the company's Paris factory produced military trucks. (They were converted for use as postal trucks after the war.) Throughout that period, many Saurer trucks continued to use wooden wheels, even though the general industry trend was toward solid rubber. Engines came in 321.7, 412.7, and 491.6ci (5.3-, 6.8-, and 8.1-liter) sizes, with outputs of 42, 52, and 62hp (31.3, 38.7, and 46.2kw).

From solid to pneumatic tires

By the end of the 1920s Saurer trucks had finally made the transition to solid rubber and then pneumatic tires, and many trucks were using diesel. In diesel use, at any rate, Saurer was a leader. It would be 10 years before other truck manufacturers began to use diesel extensively.

The 1930s saw the birth of the C-line series, which ranged in weight from 1 to 6 tons (1.016 to 6.09 tonnes). Smaller models contained four-cylinder, 169.9ci (2.8-liter), 50hp (37.2kw) diesel engines, or six-cylinder, 176ci (2.9-liter), 52hp (38.7kw) gasoline engines. When fuel supplies were scarce during World War II, the trucks came with wood-gas generators as well.

A product of World War II, and one of Saurer's most impressive achievements, was the heavy-duty M4H, used for transporting ammunition. The tractor was forward control (meaning the driver sat ahead of the front axle and by definition the engine) with four-wheel drive and a four-cylinder, 70hp (52.1kw) engine with a 1.5-ton (1.52-tonne) capacity. By the standards of the day, it was extremely powerful.

Below: *The 1930s saw the birth of the C-line series, which included one- to six-tonners.*

Specifications

Country: Switzerland

Year manufactured from: 1903

Engine: four cylinder gasoline powered

Transmission: shaft or chain driven, four-speed

Payload: up to 2.5 tons (2.54 tonnes)

Applications: delivery, military

Special features: T-head engine

MACK
🛠 **1905 USA**

MACK MANHATTAN

Mack's "Manhattan" truck was introduced in 1905. The Mack Bros. Motor Car Company was one of the first manufacturers to mount a cab directly over the engine, which increased driver visibility and maneuverability.

Above: *Early Mack trucks, up until 1910, went under the name "Manhattan." They featured chain drive.*

Specifications

Country: USA.
Year manufactured from: 1905
Engine: gasoline powered
Transmission: three-speed
Payload: two tons (2.03 tonnes)
Applications: urban delivery
Special features: cab over engine

These qualities, along with the Manhattan's patented "constant mesh" feature (which protected gears from being damaged or stripped by inexperienced drivers) and "selective" feature (which allowed drivers to immediately shift from high to low gear and vice versa without going through intermediate speeds) earned the truck an immediate following in the urban delivery business.

Jack and Gus Mack had come a long way since 1893, when they'd purchased a small carriage- and wagon-building firm in Brooklyn. The country was in the grip of depression and there were very few orders to fill. The brothers did, however, manage to establish a solid reputation for their wagon repairs, which would serve them well in years to come.

"Old Number One"

Business was slow, so Jack Mack had time to indulge his interest in the new "horseless carriages." He experimented, finally bringing out his first successful creation, a 40hp (29.8kw), 20-seater sightseeing bus, in 1900. Powered by a Mack four-cylinder engine, the handmade vehicle had a cone-type clutch and a three-speed transmission. After eight years of service, the vehicle that had come to be known affectionately as "Old Number One" was converted into a truck, and would log more than 1 million miles (1.6 million km) before it finally gave up the ghost.

In the meantime, the Mack Brothers Company began using "Manhattan" as the trade name for their motorized vehicles. In 1905, they moved their facilities to more spacious quarters in Allentown, Pennsylvania.

SAF

1907 Germany

SAF TRUCK

SAF (*Süddeutsche Automobilfabrik*) made commercial vehicles in Gaggenau, Germany. SAF engines were ahead of their time—their 1907 truck had an overhead camshaft in a four-cylinder petrol engine.

Above: *The Gaggenau truck of 1910, a cab-over-engine layout.*

Specifications

Country: Germany
Year manufactured from: 1907
Engine: gasoline powered four cylinder
Transmission: not available
Payload: 3 tons (3.04 tonnes)
Applications: multiple, variety of bodies
Special features: one of first trucks with two-pane windshield

The engine was positioned over the front axle and the driver was seated further back and high up, which made for better visibility. As for the chassis, it could be fitted with a variety of bodies. The tires were spoked and solid rubber (smaller in front), and there was actually a two-panel windshield, among the first of its kind at that time. Clearly, SAF engineers were already thinking about making the driver more comfortable.

By 1907, SAF had attracted the attention of Carl Benz, who acquired a holding in the company.

SAF responsible for trucks

The new partners coordinated their product ranges. "Benz & Cie." specialized in passenger cars and SAF assumed responsibility for the production of all commercial vehicles. The SAF name would go on to achieve great fame worldwide when the first crossing of the African continent by motor vehicle was achieved in 1909 (it took 630 days) in an SAF truck. Built to adventurer Paul Graetz's specifications, the chassis featured extra-large wheels to raise ground clearance to 14in (35cm).

AUTOCAR
🛠 **1908 USA**

AUTOCAR XVII

The original Autocar company began as the Pittsburgh Motor Car Company, founded in 1897 by brothers Louis and John Clark. Two years later, they moved the company to Pennsylvania, and renamed it Autocar.

Above: *Early Autocar trucks were of 2 tons (2.03 tonnes) payload capacity and of forward-control layout, which the company claimed provided more even weight distribution and greater load space within the overall length.*

The company began focusing on trucks in 1907, with the introduction of the Autocar Model XVII—a two-cylinder 2-ton (2.03- tonne) truck with shaft-driven axle. By 1911, the company had switched over to making trucks exclusively.

The Autocar truck chassis quickly earned the reputation of lending itself to almost every style of body, a foreshadowing of the truck's forthcoming use in multiple applications. In 1925, Autocar advertised in the *National Petroleum News* that "More even weight distribution on all four wheels is an added Autocar advantage for hauling shifting loads of gasoline and oil." In 1926, the company introduced the Autocar Model 26, a four-cylinder 5-ton (5.08-tonne) vehicle. A six-cylinder, cab-forward design was introduced in 1930.

First diesel engines

In 1935, the first diesel engines, supplied by Kenosha, were introduced. By 1940 Autocar offerings included heavy trucks with a variety of tractions, including 6x2s, 6x4s, 6x6s, and even 8x4s.

Autocar was the first truck manufacturer to build wheels with wooden instead of wire spokes, and is credited by automotive historians with several other truckmaking firsts. It was the first company to mass produce a truck with forward control (a configuration in which more than half the engine length is behind the farthest-front point of the windshield base and the steering wheel hub is in the first quarter of the vehicle length). It offered the first four-cylinder-engine forward-control truck. And it was one of the few truck manufacturers to make its own engines, transmissions, and rear axles. In addition, some historians cite Autocar trucks as being instrumental in the Allied forces' victories in both World Wars.

Advertisements appeal to patriotism

A print advertisement published in 1943 cashed in on the patriotic sentiment aroused by the Autocar at that time:

"TANKERS AWEIGH ! Massive, six-wheel tank trucks...for the Navy...Bristling Half Tracks for the Army...Rugged mobility for the Marine Corps and the Air Forces.... All essential. All for Uncle Sam. All training the men and women of Autocar to build the sturdy, dependable Autocar Truck that the world will need when war is won. So keep your pledge to the U. S. Truck Conservation Corps. Your trucks are your own, but their life belongs to the Nation."

Autocar obtained U.S. government authorization for heavy-duty hauling in 1945, a fact the company did not hesitate to boast of in print advertisements:

"HEAVY TANKAGE! For the ruts of War or the highway of Peace, Autocar Trucks are precision-built for the heaviest of heavy-duty work. Mile after mile, day after day, year after year, these famous trucks put on dependable, low-cost-per-mile performance and reduced hauling and delivery costs for the nation's leaders. Ask Shell Union Oil Corporation. They know Autocar Trucks through long years of rugged use.... Follow the leaders, for they know the way," reads an advertisement printed in 1945.

White buys controlling interest

The White Motor Car Co. purchased a controlling interest in Autocar in 1953, and White was purchased by Volvo in 1981. In 1988, Volvo GM Heavy Truck Corp. was formed as a joint venture between Volvo and General Motors Corp. At the time, Volvo GM's nameplates were WHITE GMC and Autocar. The WHITE GMC nameplate was discontinued in 1995 and Volvo GM's trucks were sold under the Volvo and Autocar nameplates, the latter continuing to appear on heavy-duty construction trucks.

In July 2001, Grand Vehicle Works Holding Corporation (GVW) announced the purchase of the AutoCar Class 8 Xpeditor truck line from Volvo. Called LCF trucks (Low Cab Forward) the trucks were used primarily used in the vocational market. Autocar celebrated its 100th birthday in 1997.

Specifications

Country: USA

Year manufactured from: 1908

Engine: two cylinder gasoline powered

Transmission: three-speed

Payload: two tons (2.03 tonnes)

Applications: multiple, body customized for buyer

Special features: Chassis lent itself to almost every style of body

Left: *After 1919 Autocar moved up the weight range with models capable of carrying 3 tons (3.04 tonnes), 4 tons (4 tonnes), and 5-ton (5.08-tonne) payloads. Forward control was the norm until 1926 when conventional trucks were added to the range.*

INTERNATIONAL HARVESTER
1909 USA

INTERNATIONAL HARVESTER COMPANY AUTO BUGGY

The International Harvester Company started when Cyrus McCormick invented the first mechanical reaper back in 1831. The invention became the foundation for the McCormick Harvester Company.

Specifications

Country: USA
Year manufactured from: 1909
Engine: gasoline powered
Transmission: not available
Payload: approx 551lb (250kg)
Applications: light delivery
Special features: frame and body made of wood

Above: *The Auto Buggy featured a frame and body made of wood. It was wider than most at the time.*

In 1847, the company built its first plant in Chicago. The International Harvester Company was incorporated in 1902, as a result of a merger between McCormick's company and four other harvest-machine manufacturers: Deering, Milwaukee, Plano and Warder, and Bushnell and Glessner. The merger was a success, with sales that year topping $52 million.

By 1906, the company was also manufacturing stationary and portable engines. A manual published by the company in 1906 includes illustrations of two trucks, one powered by steam and available in 4, 6, 8, 10, 12, 15, and 20hp (2.9, 4.4, 5.9, 7.4, 8.9, 11.1, and 14.9kw) versions, and a gasoline-driven truck available in 10, 12, 15, and 20hp (7.4, 8.9, 11.1, and 14.9kw). Both had chain drive.

Bringing goods to market

Soon after, farmers were using the company's gasoline-powered Auto Buggy to deliver their goods to market. The Auto Buggy featured a frame and body made of wood. It was wider than most vehicles so as to fit the existing wagon-wheel ruts in the roads traveled by the company's rural clientele.

RENAULT
1910 France

RENAULT TRUCK

The 1910 Renault was built for versatility. It was used for a variety of applications, including delivery and general urban transportation. The Renault story started in 1898, when Louis Renault founded the company.

Above: *This 1909 model has a sloping hood which was a characteristic of all early Renaults.*

Specifications

Country: France
Year manufactured from: 1910
Engine: gasoline powered
Transmission: not available
Payload: not available
Applications: multiple, chassis also used for taxis and busses
Special features: built for versatility

By 1900, production rose to 179 cars and by 1903, Renault had built its first heavy-duty (for the time) chassis for commercial use. The chassis earned the company a contract for supplying Paris taxis and was partially responsible for production climbing to 1,200 units in 1905. In 1907, Renault had produced a bus, and introduced it in 10- and 15-seater versions at the Ninth Paris Motor Show.

As was the case with many early 1900s truckmakers, Renault owed a great deal of its success as a heavy-duty chassis manufacturer to the war. By 1917, Renault was producing the "Diamond brand" design—the first modern tank. The company produced 30 of these tanks per day at its plant in Vernisseaux.

Award winners

By the mid-1920s, Renault tractors and 7-ton (7.1- tonne) trucks were garnering awards. The company was the first to offer a road tractor fitted with a Servo vacuum assisted brake. Renault also manufactured trucks powered by gas generators to promote the use of fuels other than gasoline.

The scheme seems to have succeeded because by the 1930s Renault was producing diesel-powered trucks. Unfortunately the Nazi occupation of France in World War II temporarily tarred the company's reputation—its trucks were extensively used by Germany.

After the war, Louis Renault was found guilty of collaboration and as a result, the company was confiscated and nationalized.

SAURER
1910 **Switzerland**

SAURER TRUCK

By 1910, the Saurer Motor Company was known worldwide for its heavy-duty trucks. In 1863 the company began to manufacture textiles and then textile machinery, and stationary gasoline engines by 1888.

Perhaps the company's board of directors had been inspired by the difficulty they were having delivering their own products, or by their experience in the manufacture of stationary engines. Whatever the reasons, by 1903 Saurer had built its first truck.

In the early 1900s, motorized transportation had still, for the most part, failed to catch on. It was largely thanks to the enterprising spirit and inventive genius of company president Hippolyt Saurer that the company saw substantial growth in spite of this, soon gaining an international reputation as a world-class heavy-duty truckmaker at a time when motor transportation was still something of a novelty.

Saurer trucks did just that, their powerful heavy-duty design rapidly gaining acceptance. In 1904, the company had already manufactured several prototypes to follow up on its first truck, including chain-driven trucks and buses with engine brakes. In 1905 Saurer saw the introduction of internal-combustion engines and carburettors. And by 1907 the company had been licensed to open another factory in Vienna. (Another was opened in Suresnes, France, in 1909.)

Used for beer

By 1910, Saurer trucks were being used on Alpine roads to transport beer (breweries were among the first Saurer customers) and

Above: *Saurer is acknowledged as one of the pioneers of truck building in Europe. As early as 1910, when this chain-drive model was built, assembly plants were being established in Germany, France, and the United States to build the trucks under license.*

postal deliveries, as fire trucks, and by the Swiss Army (which would eventually become the company's biggest customer).

Saurer trucks were being imported into the U.S. as early as 1908, and by 1910 Saurer truck buyers had a range of products to choose from. One 1910 Saurer tipper truck, purchased by Captain R. Gordon Chirnside (during a tour of Europe) for use on the family farm in Victoria, Australia (and still on the farm to this day), reportedly has a 157in (4m) wheelbase, a 3.5-ton (3.55-tonne) chassis capable of carrying 5.5 tons (5.58 tonnes), a four-speed gearbox, and a fuel consumption rate of 4.8 miles per gallon (1.7km per liter). It also had solid rubber tires that are smaller in the front than at the back. The tipping tray can be elevated by muscle power applied through two large screw jacks. Another Saurer truck, manufactured in 1910, was the Saurer-Spyker, available in an 15/22hp (11.1/16.4kw) delivery van chassis and with payloads up to 3,300lbs (1,500kg).

Merger with Mack

In 1911, Saurer and the Mack brothers formed the International Motor Company, which, as of 1912, built Hewitt trucks as well. A Swiss-built Saurer also made the first transcontinental trip in North America. The truck traveled from Los Angeles to New York City, a publicity stunt hitherto unequaled by truck manufacturers, and a foreshadowing of North America's long-haul trucking industry.

Saurer trucks were also built by Man until 1924. By 1929, the company had brought out the Saurer Type A with a Cardan transmission, which used a drive shaft with universal joints. These are common wherever a drive shaft needs to turn a corner; a drive shaft with a universal joint can freely rotate through the universal joint, and no gears are required to couple the two ends. The technology is used in vehicles to this day. Saurer gasoline-driven, 34hp (25.3kw) Acs soon followed, as did the Saurer Type B truck in 1925.

Type C introduced

The 1930s and 1940s would see the introduction of several new Saurer models, including the Type C, all-terrain trucks for the Swiss Army in 1937, and all-terrain trucks for commercial use in 1949. In 1936, Hippolyt Saurer, Saurer's driving force and the primary reason for the company's success, would pass away. But the company remained to continue to innovate in the heavy-duty commercial vehicle field, with the introduction of a proprietary motor, the Saurer BLD diesel engine in the early part of the 1950s.

Specifications

Country: Switzerland
Year manufacture from: 1910
Engine: internal combustion, gasoline powered
Transmission: four-speed, chain driven
Payload: one to five and a half tons (1.016 to 5.58 tonnes)
Applications: beer and postal deliveries
Special features: engine brakes

Below: *A Swiss-built Saurer made the first cross-continent trip in North America. The truck traveled from Los Angeles to New York, a publicity stunt that has been unequaled by truck manufacturers.*

INDIANA

🛠 **1910 USA**

INDIANA TRUCK NO. 1001

The Indiana Truck Corporation was originally founded by George C. Harwood and Charles G. Barley as the Marion Iron and Brass Bed Company in 1898, but was soon experimenting with commercial vehicles.

It wasn't long before the company brought out its first truck, the Indiana No. 1001, in 1910. One of the first trucks was sold to an O. Gordon, who sold furniture in Gas City, Indiana. The No. 1001 was surprisingly long-lived, indeed, even after Gordon later sold his truck to Gas City resident Charles Stewart, Stewart would praise it in a 1923 book of testimonials by Indiana truck owners (published by Indiana itself, of course).

"Almost six years ago I bought my old Indiana Truck from O. Gordon, who had used it for a number of years," wrote Stewart. "I am using this old truck every day, doing local and long distance hauling. This truck paid for itself the first three months of service. It is very cheap to run and I have absolutely no engine trouble. It starts easily in the morning and keeps running smoothly all day. I consider this old truck equal to any job. This old Indiana certainly is a good one, even though it is thirteen years old." Reportedly, Stewart's Indiana truck had racked up 140,000 miles (225,302km). The company would continue to produce trucks for commercial and military uses up to 1933. (Bizarrely, the company would also continue to make bedsteads and bedsprings.)

Above: *Indiana prided itself on its long-lasting trucks.*

Specifications

Country: USA

Year manufactured from: 1910

Engine: gasoline powered

Transmission: chain driven

Payload: one to five tons (1.016 to 5.08 tonnes)

Applications: delivery, military

Special features: solid rubber tires

ALCO
1910 USA

ALCO 3.5-TONNER

The American Locomotive Co. built a range of heavily constructed cab-over-engine trucks under the Alco name from 1909 to 1913. The 1910, 3.5-tonner was an early COE model.

Above: *A cab-over-engine design was universal on Alco trucks.*

Specifications

Country: USA
Year manufactured from: 1910
Engine: gasoline powered
Transmission: not available
Payload: 3.5 tons (3.55 tonnes)
Applications: multiple
Special features: cab over engine

The company started as a steam-locomotive manufacturer in 1835. But by 1909, the company had designed its first truck. Their reputation for reliable, heavy-duty trucks made the company very popular and by 1912 Alco had sold more then 1,000.

A cab-over-engine design was universal on Alco trucks, which ranged from 2 to 6 tons (2.03 to 6.09 tonnes). Maximum speed ranged from 17mph (23.7km/h) for the lighter trucks to 8mph (12.8km/h) for the heaviest. Alco trucks could carry over 3 tons (3.04 tonnes) per load. Users at the time included Singer Sewing Machine, the Gimbles Brothers, and Henry Kroger & Co.'s Monopole Whiskey.

First transcontinental trek

Alco made history in 1912 when a group of five men used one of their vehicles to make the first-ever transcontinental delivery by truck. The group set out on June 20 from New York in a 3.5-ton (3.55-tonne) model, hauling a 3-ton (3.04-tonne) payload of soap consigned to the Carlson Currier Company in Petaluma, California. At the time, there were no roads as we know them, and the journey was as much a struggle for the crew as it was for the truck. On September 12 the crew and the truck finally arrived at their destination, 4,145 miles (6,670.5km) later, and with the truck's cargo in good condition. They'd made it in 91 days.

RANDOLPH

🛠 **1910 USA**

RANDOLPH MODEL C-2

The Randolph Model C-2 had a horizontally-opposed two-cylinder gasoline-powered engine. It was chain driven, with a three-speed transmission and solid rubber tires on wooden spoke wheels.

Randolph trucks were built by the Randolph Motor Car Company, which was reportedly founded by a garage owner in 1908 in Chicago. The company's first trucks were light delivery vans or trucks, and were powered, as in Model C-2, by two-cylinder engines. But the 1- ton (1.016-tonne) Model 14 would have a four-cylinder 21hp

Above: *The short-lived Randolph was only produced from 1908 to 1913.*

(15.6kw) engine, and later Randolph trucks would grow to have a capacity of as much as 4 tons (4.06 tonnes). These would be nicknamed the "Strenuous Randolphs" in print advertisements of the day, and due to the increased heat produced by the more powerful engine, they would require two radiators (in side panels under the driver's seat) and a large centrifugal pump to circulate water through them. The increased weight would also require single-elliptic springs to be used in front and full-elliptic springs to be used in the rear.

Company history

As for the Randolph Company's history, it is somewhat obscured by the sheer number of times it was sold in its short life (1908–13). It was purchased by General Motors and moved to Flint, Michigan, in 1908, and in 1912 it was sold again. Some historians say it was sold back to the original owner (whose name has been lost in the mists of history), at which time its name was changed to the Randolph Motor Truck Company, obviously with the intention of focusing production on commercial vehicles. Another report states the company was actually sold to the Sterling truck company, and yet another report has it that Randolph was renting factory space from the General Motors company and was thrown out when the factory was sold to Sterling Trucks.

Whatever the case, historians generally agree that Randolph stopped making trucks in 1913. The Randolph name, however, has lived on and examples of Randolph trucks are still on display in some American museums.

Below: *The Randolph Model C-2 could be used for various applications. It is seen here as a delivery truck.*

Specifications

Country: USA
Year manufactured from: 1910
Engine: horizontally opposed two cylinder gasoline powered
Transmission: chain driven, three-speeds
Payload: 1102.3lb (500kg)
Applications: multiple
Special features: steering wheel on right-hand side

KISSEL KAR

1910 USA

KISSEL KAR FIVE-TONNER

The Kissel Kar Company was founded in 1906 by brothers William and George Kissel. The brothers were reportedly inspired by their family's own "horseless carriage", although the make of that vehicle is unknown.

Specifications

Country: Germany
Year manufactured from: 1910
Engine: four cylinder
Transmission: belt or chain driven
Payload: 0.5–5 tons (0.5 to 4.08 tonnes)
Applications: delivery, military
Special features: optional removable closed cab

The low-slung "Gold Bug" speedster was especially popular with celebrities of the day, who even traveled all the way to Hartford themselves to pick up their new automobiles. But cars weren't all the brothers would build. Kissel would also build hearses, coaches, fire trucks, taxis, and other styles of vehicles. It was only a matter of time before they started manufacturing trucks.

The first Kissel heavy-duty truck was produced in 1910. It would have a four-cylinder diesel engine (made by one of two local enginemakers at that time: Wisconsin or Waukesha), attached to a passenger car chassis. Also available would be an optional (and removable) winter driver's cab, with plate-glass windows, which were a novelty for the nascent truck industry. The Kissel trucks were strong but they were also unusually stylish for their time. In fact, one of the first large orders for the trucks was placed by a circus.

Above: *Early Kissel heavy trucks had Wisconsin or Waukesha gasoline engines and some featured a differential lock to prevent wheel slip on slippery roads.*

FEDERAL
1910 USA

FEDERAL TRUCK

In its newsprint advertisements, the Federal Motor Truck Company of Detroit, Michigan, proclaimed itself the "largest manufacturer of 4½ and 3½ ton (4.57 and 3.55 tonne) worm drive motor trucks exclusively".

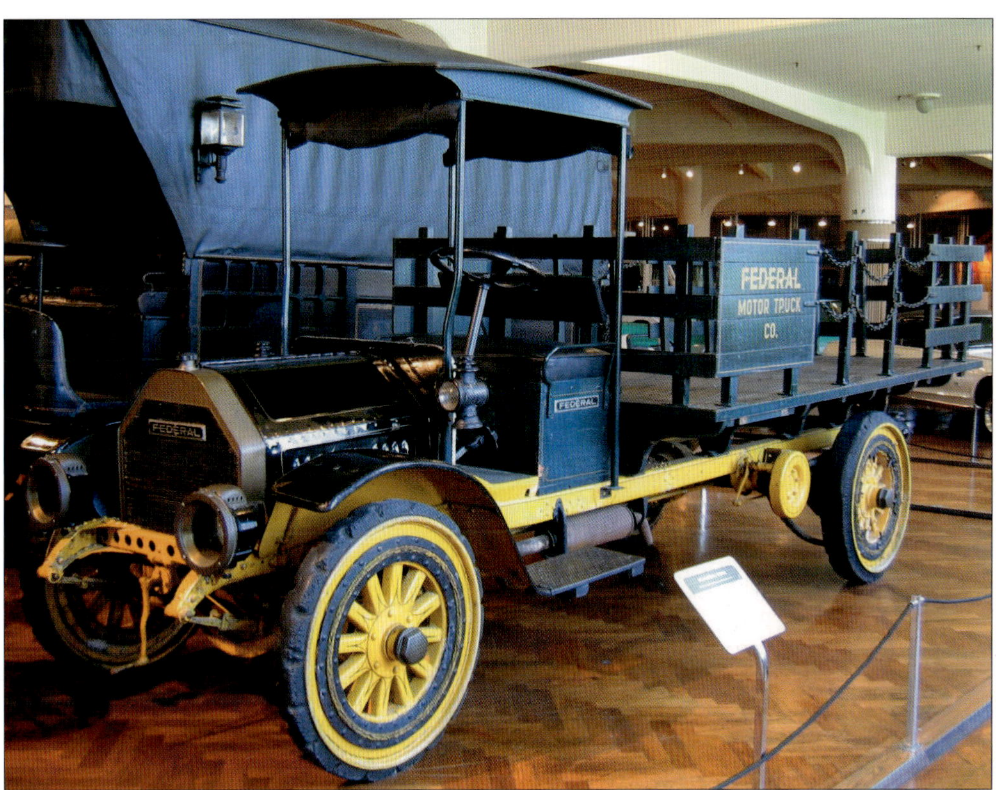

Right: *What originated as the Bailey Motor Truck Co became the Federal Motor Truck Co and the first chain drive 1-tonner of 1910 was based on a Bailey prototype. Power came from a 4-cylinder gasoline engine.*

Specifications

Country: USA
Year manufactured from: 1910
Engine: gasoline powered
Transmission: belt drive
Payload: one ton (1.016 tonne)
Applications: haulage, agricultural, delivery, municipal
Special features: by 1920s Federal trucks were being exported overseas

At that time, Federal mainly manufactured trucks that could carry a 1-ton (1.016-tonne) payload. The 1910 design was highly conventional, and looked very much like a car, with its radiator and gasoline engine in the front and its driver sitting behind it under a metal awning.

It had spoked steel wheels with solid rubber tires, and the chassis had two frames rails, which appear to be as close together as a car's. (The wheels, however, were set out from the frame rails, making it easier for them to fit into the ruts that were typical of many American roads at that time.) The truck could reach speeds of up to 20mph (32km/h).

Ford an early convert

Henry Ford, of the Ford Motor Company, would buy the very first model, fitted out with a wood-slat box on top of a flatbed, for use on his farm. This vehicle survives in the Henry Ford Museum in Dearborn, Michigan, to this day. Ford was one of the truck's earliest converts (commercial motor vehicles were, at that time, still competing with horse-drawn carriages). In fact, many motor-truck advertisements of the time pointed out how motorized transportation was a more cost-effective haulage method than horse-drawn carriages. Federal was not alone in trying to convince the potential truck buyers of this.

COMMER

1911 UK

COMMER TRUCK

"What do you really know about your delivery costs?" reads a 1911 print advertisement for the Commer truck. The ad goes on to profess the Commer's superiority in terms of cost-per-ton mile.

Above: *Early Commer trucks like this 1911 example were created around a revolutionary new preselective gearbox known as the Linley gearbox.*

Commer got its start as Commercial Cars Limited in London, in the early 1900s, but soon erected its first plant in Luton, in Bedfordshire. In 1904, the company brought out its first light-duty truck. It proved popular and the company quickly produced a range of light and medium-duty commercial vehicles, like the Commer pictured above, many of which would be sold outside England well before World War I. (Among the company's exports was a commercial vehicle with straked wheels for use in South America, another with a power-loading gear for New Zealand, and another with steel tires for transporting oil drums in Siberia.) In 1911, the British-made, gasoline-driven Commer, with chain drive, spoked wheels, and (at first) solid rubber tires, with an optional hood, was available in 2½-, 3½-, 4½-, and 5½-ton (2.54-, 3.55-, 4.57-, and 5.80 tonne) chassis versions. With a full load the 4.5-ton (4.57 tonne) model could run nearly six miles to the gallon (2.1km per liter) on a straight road. The truck was also being exported to North America.

Name change

During World War I, the company focused its manufacturing efforts on military trucks. No fewer than 3,000 Commer military trucks had been built by 1919. Then in 1926, Commercial Cars Ltd. partnered with Humber Ltd. and the company name was officially changed to Commer (although its trucks had long been referred to by that name). Two years later both companies were absorbed into the Rootes Group.

Multipurpose vehicles

One of the most famous Commer trucks during the 1930s was a light normal- or forward-control van. Just before the World War II, the "Superpoise" was lauched. This Q Model was very popular in England. The Superpoise chassis was built as 15cwt (0.75 tonnes) up to 6 tons (6.1 tonnes). The truck had a fully enclosed cab and various Humber or Commer gasoline engines from four to six cylinders. Many of these 15-cwt (Q15), 2-ton (Q2), and 4/5-ton (Q4) trucks were adapted for military purposes during the war with a great variety of bodies, such as explosives vans, general service vehicles, troop loaders, or fixed container workshops. The military Commer also received dual rear tires, a special oil bath cleaner, a larger radiator, and a tubular guard positioned at the front of the grille.

"Screamers"

Later, Rootes Motor Group would build a horizontally-opposed, two-stroke diesel engine for its trucks that used a system of rockers to transmit crankshaft motion to the pistons. The engine, which made a peculiar "screaming" sound, especially when pulling uphill in low gear, would be widely used in Commer trucks. The trucks would be aptly nicknamed "Screamers" as a result.

Below: *Commer was first established in southwest London in the early 1900s.*

Specifications

Country: UK
Year manufactured from: 1911
Engine: gasoline powered
Transmission: chain drive
Payload: up to 5½ tons (5.08 tonnes)
Applications: multiple
Special features: optional hood

MACK

1914 USA

MACK AB SERIES

The Mack AB medium-duty truck was the company's first standardized, high-volume model series, introduced in 1914. The first ABs had chain drive. Dual-reduction drive would be offered as a replacement in 1920.

Above: *Sometimes known as the "Baby Mack," the AB could haul up to 3 tons (3.04 tonnes). Most ABs had chain drive but this example features the alternative shaft drive.*

The original ABs were available in versions of 1½ to 3 tons (1.52 to 3.04 tonnes). Their engines used a cross-shaft for driving the magneto and water pump. They also had large inspection ports, which could easily be removed to inspect the connecting rod bearings (the Mack brothers knew how important ease of maintenance was). Early AB engine models had a governor that prevented the truck from reaching more than 16mph (25km/h), the top speed for trucks using solid rubber tires at that time. The Mack ABs had a new, engine-forward design, lauded by some as an improvement on cab-overs. More than 50,000 AB series trucks would be made through 1937.

Mack would also build an armored car on an AB chassis for the New York National Guard (in 1916). But by then, many of the Mack brothers' executives who had joined the company over the years, had left, thanks to a top-level management dispute, brought about by a merger intended to help the company expand.

By 1909, Mack had already brought out its heavy-duty "Senior" line of trucks and its smaller, lighter-duty truck line, dubbed "Junior." The Junior model was a 1½-ton (1.52-tonne) truck, with left-hand steering and worm drive, while the Mack Senior truck was right-hand steering. The company had also produced the first motorized hook-and-ladder fire truck for the city of Morristown, New Jersey (1910).

Mack trucks also began sporting the "Mack" nameplate (instead of Manhattan, which was originally used to distinguish the brothers' horse-carriage business from their motor-vehicle business), and a new company, The Manhattan Motor Truck Company, was incorporated in Massachusetts to operate several dealerships in that state.

600 units per year

By 1911 Mack was well established in North America as a manufacturer of heavy-duty trucks, making 600 units a year. But the company just couldn't produce enough to keep up with demand. The company needed more money to expand and financier J.P. Morgan organized the merger of the Mack Company with the Saurer Motor Company (with a factory in Plainfield, New Jersey) to form the International Motor Company. The two companies continued to function as distinct organizations, but the selling and servicing of Mack and Saurer trucks was combined as a function of the holding company (International). The company produced 2,000 trucks in 1911 alone.

Mack goes conventional

In 1912, the International Motor Company absorbed the Hewitt Motor Company as well. (Hewitt was a builder of highly engineered motor trucks, based in New York City.) John and Joseph Mack, who'd been directors of the International Motor Company left, but Jack Mack remained as vice president. The impact was immediate. The entire company's production fell sharply, dropping to 1,150 units by the end of 1912.

The company would eventually drop its other lines and revert back to Mack Trucks. By the end of World War I, the name of the company was changed to the Mack-International Motor Truck Corporation, and Saurer truck production was discontinued in the United States. But Jack Mack would be gone by then, unhappy with the changes.

Specifications

Country: USA
Year manufactured from: 1914
Engine: gasoline powered
Transmission: worm drive, later dual reduction
Payload: one and a half to three tons (1.52 to 3.04 tonnes)
Applications: delivery, military
Special features: engine-forward design

Left: *The Mack AB used a cross-shaft for driving the magneto and water pump. It also had large inspection ports, which could easily be removed to inspect the connecting rod bearings.*

chapter 3

The War Years

Cost-effectiveness was the early battle cry of truck manufacturers. "If your hauls go more than 30 miles per day we can prove the Commer Truck more economical than horse and wagon trucking," was just one example of the general marketing strategy of the period. After all, travel distances were increasing and it was actually becoming quicker and more cost-effective to use motorized vehicles.

Above: *It was realized that a standard truck was needed to make truck maintenance and repair easier for the Allied forces in Europe and so the Liberty truck was designed.*

Left: *The Pierce Arrow R-5 was used extensively during World War I by all the Allied armies.*

Road systems throughout Europe, Asia, and North America had improved somewhat, but they still left much to be desired. Truck durability was vital, and never did durability and the ability to navigate rutted, muddy, and off-road surfaces prove more important to the early adoption of trucks than during World War I, when they were first used and produced on a massive scale. The war may not have been kind to humans, but it did give trucks the boost they needed, both in the popular imagination and in terms of rapid technological development.

Meanwhile, in the United States production exploded. American truckmakers were more than happy to take the opportunity to export their product to meet the needs of war-torn countries. Still, there would be casualties among truck manufacturers. They would come after the war, in countries like Germany, whose industrial might was badly damaged by occupation, and in the United States, when truckmakers who had invested in mass production of military trucks in wartime were left holding the bag as their own products flooded the consumer market.

World War I was a boon to many truck manufacturers, but it would also be the end of others. Whether individual truckmakers survived or perished, they had all earned the truck a permanent place on the roads.

GERSIX

1915 USA

GERSIX MODEL G

Brothers Louis and Edward Gerlinger founded the Gerlinger Motor Car Company in 1912, selling cars built by other manufacturers. It wouldn't be until 1915 that they would get into the truck business themselves.

Above: *The Gersix was the forerunner of the famous Kenworth. This is an early shaft-drive "G" model for 3.5-ton (3.55-tonne) payload.*

The first Gersix truck would be the end product of a "make-work" project. Business was slow at the repair shop that the Gerlingers operated as a sideline, and they really wanted to keep their mechanics busy, so the brothers decided to have their mechanics build trucks between the repair jobs, out of the spare parts of trucks built by other manufacturers.

It perhaps wasn't the wisest business choice at the time, with truck parts already beginning to become scarce as a result of the war in Europe, and a lot of bigger manufacturers adopting the same strategy. But the Pacific Northwest was potentially a niche market. There were no other truck makers specializing in making vehicles that could handle the rough terrain and extremely steep grades of the region, not to mention the logging loads that were a basic industry in those parts, and Gerlinger mechanics had more than enough time to scrounge for parts.

The first truck was the 2.5- ton (2.54-tonne) chain-driven Gersix Model G, without a cab or body. The "Ger" stood for Gerlinger and the "six" stood for the truck's six-cylinder Continental engine, an anomaly when most manufacturers were making trucks with four-cylinder engines. The first Gersix to be built was sold to a brick hauler based in the Portland area.

Gersix Model Gs

Two more Model Gs would be produced a year later. But parts were getting even harder to find. When the Gersix's axle supply was cut off, the company would be forced to make its own. The company's first truck production line wasn't much to look at. One mechanic would later describe it as nothing more than "two chalk marks on the floor" where workmen put the axles, then filled in the truck in between.

Even so, the truck was specially designed for the extra-heavy multiple trailer loads and axle-snapping terrain of the Pacific Northwest—a fact that would eventually win the Gersix local popularity. It didn't happen overnight, however. Owners of logging companies were understandably reluctant to spend their hard-earned cash on an unknown brand of truck, so the Gerlinger brothers had to haul loads from town to town to prove what the truck would do.

Niche market

In 1917, the company would move to Tacoma, Washington, and begin to sell Federal trucks in addition to its own. While orders were coming in, the Gerlinger brothers still didn't have the means to produce more than two trucks per month. So they sold their company to a group of satellite investors, including a man named Edgar Worthington.

There were still problems to be overcome, however. Despite the injection of new funds, material shortages caused by the war were beginning to hurt production. Iron shortages would force the company to reduce the number of cylinders in its engines to four. Gersix trucks would have to be equipped with four-cylinder Buda engines, which would prove adequate but would take the"six" out of "Gersix."

Still, the postwar period proved brighter for Gersix trucks than some other companies. While wartime government orders had ceased, truck sales in the Washington area were soaring for Gersix, which in 1922 sold 523 trucks, almost all of them in the Pacific Northwest.

Harry Kent

The same year also saw the arrival of Harry Kent, who joined the company as an engineer. With Kent, whose expertise was in cutting costs and meeting budgets, the company would see further success. By 1923 the company had changed its name to Kenworth Motor Truck Company ("Ken" for Kent and "Worth" for Worthington.) Kenworth would also open a new factory in Seattle and produce 78 new trucks that year.

Specifications

Country: USA
Year manufactured from: 1915
Engine: six cylinder Continental
Transmission: chain driven
Payload: two and a half tons (2.54 tonnes)
Applications: logging, delivery
Special features: designed especially for Pacific Northwest

Below: *The Gerlinger Bros. output of two trucks per month failed to meet demand and they sold the company in 1922 to the founder of the Kenworth Motor Truck Co.*

MACK
1916 USA

MACK AC "BULLDOG"

The Mack AC truck was introduced in 1916. With its chain-drive rear axle, the truck soon earned a reputation for reliability and durability. It saw continued success, remaining in production for 24 years.

Even before World War I, Mack had established a good international reputation. The company had already built many British military trucks, as well as an armored car on an AB chassis for the New York National Guard.

Chief engineer Edward Hewitt was credited with designing Mack's AC model, which had 18 patents to cover all of its innovations. The truck's radiator was placed behind its engine, and air was drawn from the hood, forced through the radiator, and then sent through screening at the front. This unique design was believed to prevent the damage to exposed radiators that was apparently being caused, accidentally or not, by antitruck lobbyists. (It was led by U.S. trade unionists who feared, justifiably as it turns out, that motorized vehicles would put horse-drawn transportation out of business.)

Another innovation was the positioning of the steering wheel at a 45-degree angle, which would prove to be considerably more comfortable for the driver. As for the engine, it was rated at 74bhp (55.18kw). (The acronym "hp" stands for "Brake Horsepower" and refers to the actual or useful horsepower of an engine, usually determined from the

Above: *The Mack AC "Bulldog" had a formidable reputation for toughness. British soldiers in World War I were so impressed by its tenacity that they nicknamed it "Bulldog." Mack subsequently adopted the quintessentially British breed as its company emblem.*

force exerted on a friction brake or dynamometer connected to the drive shaft.) The press chrome-nickel steel frame rails were heat-treated for durability, as were the pair-cast cylinders. Lightweight aluminum was used for the radiator tank, the transfer case, the timing cover, and the engine crankcase. The front axle was made of dropforged alloy steel to make the AC tougher.

Selective transmission

The AC had a selective three-speed transmission and a clutch brake. It also had an all-steel cab with an optional metal roof—a substantial improvement on the more common exposed wooden seats of the day. The design proved extremely popular. During World War I, Mack delivered approximately 4500 AC model trucks of 3.5-ton (3.55-tonne), 5.5-ton (5.58-tonne), and 7.5-ton (7.62-tonne) capacity for the government of the United States, which had adopted the truck as a standard military vehicle. These went overseas to France with American doughboys.

Mack also produced and delivered more than 2,000 units to Great Britain during the same period. British soldiers reportedly referred to Macks as "bulldogs," calling out "Aye, send in the Mack Bulldogs!" when faced with a difficult transportation problem. British mechanics and soldiers apparently believed the trucks had the tenacity of a bulldog. As the bulldog was the symbol of Great Britain at that time, this was high praise indeed. The Americans needless to say, shared the opinion of the British, at least according to company historians. Of course, the nickname could also easily have been due to the appearance of the trucks, with their snub noses. Whatever the case, the nickname stuck. AC and Mack trucks in general would eventually became widely known as "the Bulldog Macks," and by 1922 the company adopted the Bulldog as its corporate symbol.

Bulldog ornament

As for bulldog hood ornament, it wouldn't come along until much later. First, a picture of a bulldog (first drawn in 1921) would appear on a sheet-metal plate riveted to each side of the cab, showing the bulldog chewing up a book entitled "Hauling Costs," with the name "Mack" on his collar and "International Motor Company" underneath. (The name of the company, which had been changed to International Motor Company after a merger with two other truckmakers, would be changed back to Mack in 1922 to prevent the company from being confused with another US truckmaker, the International Harvester Company.)

Left: *In frontal appearance the Mack AC, with its radiator-behind engine layout, bore an uncanny resemblance to the French Renault. Ironically Mack was eventually absorbed into Renault Véhicules Industriels in 1990.*

Specifications

Country: USA
Year manufactured from: 1916
Engine: gasoline powered
Transmission: three-speed selective with clutch brake
Payload: one to seven and a half tons (1.016 to 7.63 tonnes)
Applications: delivery, military
Special features: radiator behind engine

Below: *Even before World War I, Mack had established an international reputation.*

GARFORD
1916 USA

GARFORD TWO-TON 4X2 TRACTOR

The Garford Company had its start in 1909, when the Federal Manufacturing Co., which made automotive components, was renamed. The first vehicles produced under the Garford name would be taxicabs.

Above: *Starting in 1915, Garford focused exclusively on truck manufacturing until they were taken over by Relay Motors in 1927.*

Specifications

Country: USA
Year manufactured from: 1916
Engine: gasoline powered four cylinder L-head
Transmission: four-speeds, dual chain drive
Payload: five tons (5.08 tonnes)
Applications: multiple
Special features: cab before engine

The very first Garford trucks were cab-before-engine models with a 5-ton (5.08-tonne) payload. They had four-cylinder L-head gasoline engines, four-speed transmission, and dual chain drive connected to the rear axle.

The company started to focus exclusively on truck manufacturer as the Garford Motor Truck Company in 1915. In 1917, the company introduced three new truck tractors with 4.5-, 7-, and 10-ton (4.57-, 7.1, and 10.16-tonne) payload capacities. Garford trucks would be used by the military during World War I. Some of them would even be half-trucks. The company would also build about 10,000 Liberty B trucks commissioned by the United States government for military use.

Format change

The biggest change came in the 1920s, when the company changed from cab-over-engine to conventional models ranging from 1 to 5 tons (1.016 to 5.08 tonnes) in payload capacity. All of these were powered by Buda four-cylinder engines with four-speed transmissions and chain or worm drive. They soon had disk wheels.

Above: *Garford built a range of conventional trucks with a payload capacity from 1 to 5 tons (1.016 to 5.08 tonnes), powered by Buda four-cylinder gasoline engines.*

WINTHER
1917 USA

WINTHER FOUR-WHEEL DRIVE TRUCK

Above: *Winther tapped into the growing demand for trucks with four-wheel drive for military use in World War I.*

The Winther Motor Truck Corporation was started by Martin P. Winther in 1917. He was previously an engineer at the Thomas B. Jeffrey Company, but left the company when it was sold, preferring to strike out on his own.

Right: *The Winther Motor Truck Corporation appears to have succeeded in its efforts to build at least a small range of vehicles suited to municipal applications.*

Specifications

Country: USA
Year manufactured from: 1917
Engine: gasoline powered
Transmission: chain drive
Payload: not available
Applications: multiple, urban, and municipal
Special features: excellent traction due to four wheel drive

His timing would prove ideal. There was a dire need for more U.S. military trucks, especially four-wheel drive types like those initially designed and developed by Jeffrey's company, for the war in Europe.

Right from the start Winther made trucks only, and like most manufacturers at that time, benefited from increased demand during the war. In fact, the company was among those consulted when the U.S. Army decided to come up with the design for the standardized Liberty B truck. Winther would also produce Liberty trucks, as well as its own models for use overseas.

Meanwhile, at home the company continued to produce four-wheel drive models for commercial and municipal use. Again, thanks to the war, there was a growing market for four-wheel drive trucks in North America as well as in war-torn Europe. Business owners had slowly begun to recognize the advantages of four-wheel drive when it came to navigating both city streets and rough and muddy terrain.

Used by municipalities

Several large municipalities depended on Winther trucks for garbage disposal, including Minneapolis and Sioux City. Unfortunately, Winther would suffer when World War I ended, and soon fade from existence. The company mounted a last effort as the "Kenosha Fire Engine and Truck Co." before closing.

OSHKOSH
1917 USA

OSHKOSH ALL-WHEEL DRIVE CHASSIS "OLD BETSY"

Thomas B. Jeffrey may have been one of the first to develop a four-wheel drive truck (later known as the Nash Quad), but he wouldn't be alone in that regard for long.

The Wisconsin Duplex Auto Company (which would adopt the Oshkosh name later) was founded in May 1917. Its first four-wheel drive truck, which would come to be known as "Old Betsy," rolled off the production line soon afterward. Not much later, the company relocated to Oshkosh, Wisconsin, renaming itself after its brand new hometown.

All-wheel advantages

The all-wheel drive truck was powered with a four-cylinder LeRoi engine. A three-speed gearbox was also fitted. It had a transfer case, with the patented automatic positive-locking center differential. As a "prototype," the vehicle demonstrated all the advantages of the all-wheel drive system. It was a success—the truck sold after just a few weeks.

Above: *In 1917 the company's first four-wheel drive truck was introduced, this became known as "Old Betsy."*

Model A truck

By 1920, the Oshkosh Motor Truck Manufacturing Company brought out the Model A, its first series-production truck. It would also be among the first trucks to be optionally equipped with pneumatic tires. It was available with a variety of bodies, and could be fitted with an enclosed cab. It was shaft-driven and the driver rode behind the engine, which was positioned over the front axle. The Model A engine was rated at 53.6 kw (72 hp), a very high horsepower at this time and it looked very modern with an electric starter, an electric horn, two headlights etc.

Model H truck

In 1925, a new Model appeared briefly. It was the Model H, with double reduction axles, a great step in the automotive industry. This truck was designed for severe duties like snow removal or road construction. Oshkosh sold the truck fully kitted out for snow removal, with all the accessories.

Soon the company would move to a new, larger facility on the south side of Oshkosh to meet the growing demand for its trucks, and the company would introduce the Model B and Model F to meet the demand.

Model TR

In 1933, Oshkosh built its first articulated earth mover called the Model TR. This 4 x 4 had an open cab, four-wheel steering, and a bottom-dump semitrailer. With its Goodyear tires (the largest made to that time), the TR was a great success and more than 100 were sent to England. Two years later, Oshkosh introduced a new cab with the J series, which featured a one piece windshield. Just before the war, the famous W series was launched. These stayed in production until the sixties and the first series were sold for military applications at the beginning of the war.

Below: *By 1920, the company brought out the Model A—its first series production truck.*

Specifications

Country: USA
Year manufactured from: 1917
Engine: four cylinder side-valves engine
Transmission: three-speed gearbox with transfer case to propshaft
Payload: one ton (1.016 tonne)
Applications: off road
Special features: four-wheel drive with automatic locking center differential

FORD TT
1917 USA

FORD TT ONE-TONNER

Henry Ford built his first experimental commercial vehicle—a panel truck—in 1899. But it wasn't until 1917 that the Ford Motor Company brought out the TT one-tonner—basically a beefed-up Model T car.

Above: *Henry Ford's legendary Model T was joined by a heavier model, the TT 1-tonner, in 1917. The simple rugged low-cost workhorse took the truck market by storm.*

A machinist's apprentice at the tender age of 16 in 1863, Ford's engineering skills had become evident from an early age. He soon became chief engineer for the Edison Illuminating Company in Detroit, Michigan. By 1903 Ford and a group of investors, who believed in his genius, had founded the Ford Motor Company.

First Ford truck

The first commercial vehicle produced by the company was the Model E delivery wagon (a later version of the panel truck mentioned above). It proved useful for city deliveries and Ford Motors would continue to manufacture it through 1911. By then Ford had become known mainly for manufacturing light trucks and cars. Most famous of all, of course, was Ford's Model T car, otherwise known as the "Tin Lizzy."

But Henry Ford soon gained renown as an industrialist as well. He became the first automotive manufacturer (and the first North American manufacturer) to introduce assembly-line production to his factory. He was also the originator of salaries of $5 per day for workers (which was considered a good wage at that time) as well as the creator of the 40-hour working week.

Ford would continue to develop trucks. The Model TT truck, produced in 1917, was Ford's first venture into the big-rig market. With its solid rubber tires, wood-spoked wheels, and canvas tarpaulin, the first Model TT wasn't much to look at. But later versions, with carved wood-paneled bodies, solid roofs, electric lights, pneumatic tires, and glass windshields, would substantially improve the TT's overall looks and saleability. The first TT was powered by a 177ci (2898cc) engine (positioned over the front axle and in front of the driver, "conventional" style) with three-speed transmission. The truck chassis, though modeled on the Model T, had a longer and sturdier frame. The wheelbase was also 2ft (61cm) longer and the Model TT had a stiffer rear suspension than the Model T did.

Ford creates modern assembly line

As he'd done with his other vehicles, Ford immediately put his assembly-line magic to work for the Ford Model TT. No fewer than 25 Model TT truck chassis rolled off the production line each day. By 1922, the number of trucks produced by Ford per day had grown to 193,294 units, a testimony to Ford's ability to survive the war's end, thanks to the fact the company continued to produce civilian vehicles throughout the war. In short, while the war in Europe had ultimately meant the end of companies that had overinvested in wartime truck production, it had only helped increase Ford's truck sales abroad and at home.

The Depression years that followed had their effect, with Ford truck sales falling just like everyone else's. Still, because of Ford's light-truck business and the affordability of its vehicles, its sales remained healthier than most. What finally brought full recovery to the company was Ford's introduction of the Flathead (also known as an L-head) V-8 engine in 1932. The engine would be offered in all the company's truck models as well as its cars.

Introduction of V-8

Ford engineers cut corners by casting the engine in a single piece rather than several. By doing so they also reduced the number of engine parts, which increased durability (but also made it harder to repair). The end result was that the V-8 engine boosted Ford truck and car sales and reestablished the company as the industry leader.

Inset: *Simplicity was the keynote of Ford's design, exemplified by its single transverse front spring. The TT was inexpensive to buy and inexpensive to maintain.*

Specifications

Country: USA
Year manufactured from: 1917
Engine: gasoline powered
Transmission: three-speed, chain driven
Payload: one ton (1.016 tonne)
Applications: delivery
Special features: later versions included electric lights, pneumatic tires, etc.

Below: *One of the TT's strengths was its adaptability to a host of different special duties, such as this dump truck for small consignments of building materials.*

PIERCE
1915 USA

PIERCE ARROW R5

Although renowned for its luxurious cars, the Pierce-Arrow Company built commercial vehicles too. The most famous of them was a 5-ton (5.08-tonne) truck that established a reputation for reliability and quality.

Above: *The Pierce Arrow R5 was in big demand for military service during World War 1.*

Specifications

Country: USA
Year manufactured from: 1915
Engine: four cylinders, side valve, 40hp (29.8 kW)
Transmission: four-speed transmission, low gear reduction
Payload: five tons (5.08 tonnes)
Applications: commercial, delivery, dump truck
Special features: worm drive

Above: *After production had already reached 7000 in 1917, Pierce Arrow was contracted to build another 1000 Class B Liberty trucks.*

In 1896, George N. Pierce founded a company that, along with bicycles and motorbikes, soon began producing motorcars powered by the famous DeDion engine. In 1906, the growing company moved to Elmwood, and in 1907 the first commercial truck was completed. In reference to the bicycle and motorbike nameplates (an arrow), Pierce chose the Pierce-Arrow name in 1910. The first truck series was a 5-ton (5.08-tonne) model named X-1. At this time, Kerr Thomas (from Saurer), David Fergusson (Leyland), and John Younger (Dennis) joined the company and began to work on a complete range of trucks from 3½ ton (3.55-tonne) to 5 ton (5.08-tonne). One of the most popular trucks after 1915 was the R-5, a new worm drive 5-tonner. It was extensively used during World War I by all the Allied armies. This commercial vehicle remained unchanged until the end of the twenties except in its designation: The R-5 became the R-9 in 1918 and the R-10 in 1920, with some improvements. Over the following years, the Pierce-Arrow frames were used by many commercial body manufacturers for firetrucks, ambulances and police cars. But in 1932, because of the global economic crisis, the company decided to stop truck production.

LIBERTY
1918 USA

LIBERTY CLASS B

The Liberty Class B was designed by a committee of American military officials, members of the American Society of Automotive Engineers, and volunteers from 45 American manufacturers.

Above: *The need to standardize military trucks to simplify spare-part supplies and maintenance led to the Liberty, which was built by a number of U.S. manufacturers starting in 1918.*

Specifications

Country: USA
Year manufactured from: 1918
Engine: gasoline powered
Transmission: four-speed transmission, low gear reduction
Payload: three tons (3.04 tonnes)
Applications: military
Special features: produced by multiple manufacturers for military

The Liberty concept was born, like most other inventions, out of necessity. The Allied forces in Europe were having a hard time maintaining and repairing their trucks during World War I. U.S. military officials soon realized that military trucks made of standardized and interchangeable parts would improve their situation considerably.

Standard design

Specs for the standard Liberty Class B, 4x2, 3-ton (3.04-tonne) truck were submitted to U.S. Army officials in July 1917, and a design committee was formed. By October 1917, two prototype trucks had been built and were available for testing. (The engine block was made by Continental, its cylinder heads by Waukesha, and its pistons by Hercules.) The trials proved so successful that by mid-November, the U.S. government had already signed contracts for the manufacture of the Liberty B's standardized components.

The final Standard B 5-ton (5.08-tonne) would be 21ft 9in (6.6m) long by 7ft (2.1m) wide and 6ft 3in (1.9m) high. It would have four forward and one reverse gear, a 160.5-in (4.07-m) wheelbase, and front tires that were smaller than the rear ones. The truck's L-head engine would measure 425ci (255cc) and run four cycles, with 52bhp (38.7kw). Production of the Standard B Liberty truck would begin in April 1918.

WHITE
1918 USA

WHITE FIVE-TONNER

The White 5-ton (5.08-tonne) chassis could be fitted with any number of bodies for a wide variety of purposes. It was a common enough sight on North American streets by 1918.

The White Company had come a long way by 1916. Originally experimenting with the manufacture of steam-powered delivery wagons, what was then known as the White Sewing Machine Company of Ohio had been asked by the U.S. Army in 1903 to design a motor-driven ambulance that would serve as a test vehicle.

By 1911, the army had realized the potential military applications of trucks and had begun initial testing for standard army trucks. Two vehicles were initially purchased for the test, including a 1-ton (1.016-tonne) gasoline-driven shaft-drive vehicle made by the White Company (whose main business still consisted of making sewing machines). The test results were so positive that the army decided to purchase a White truck that year. And one year later, another White truck was part of a trial 1,500-mile (2,414-km) trek, from Washington, D.C., to Fort Benjamin Harrison, Indiana. Unfortunately, this test proved less successful. The self-starter had yet to be invented—drivers used a magneto-ignition that required hand-cranking, and the brake system was purely mechanical, which meant a great deal of force had to be exerted by the driver's foot.

Above: *White 5-ton trucks (5.08-tonne) were powered by a 323.4ci (5.3 liter) gasoline engine. Some of the later 5-tonners, in 1918, had more powerful engines of 360ci (5.9-liter) capacity.*

Specifications

Country: USA

Year manufactured from: 1918

Engine: gasoline powered

Transmission: bevel and chain drive

Payload: five tons (5.08 tonnes)

Applications: multiple, delivery, military, etc.

Special features: could be fitted with different bodies

Left: *In this press advertisement of 1918 White emphasized the important role played by their trucks in long-distance haulage although the illustration might be a little fanciful.*

More testing

Still, yet another trial was conducted. White trucks were among the trucks being tested, both 1-ton and 3-ton (1.016-tonne and 3.04-tonne) models, the former proving less reliable than the latter, because trucks were required to travel at a walking pace alongside the troops, causing overheating problems. The heavier vehicles did better because the troops tended to overload the trucks. By 1919 White trucks had clearly made a name for themselves. That same year, the army organized yet another convoy from Washington, D.C, to San Francisco, and cheering crowds greeted the trucks on their arrival.

Bottom: *By 1919 White trucks had clearly made a name for themselves both within the army and as commercial vehicles.*

NASH QUAD
⚒ **1919 USA**

NASH QUAD

In the beginning, trucks had to prove their worth in dollars and cents. Still, there were those who believed trucks could do it, and one of these was Thomas B. Jeffrey, the inventor of the Jeffrey Quad.

Above: *The Nash Quad was unusual in having four-wheel steering as well as four-wheel drive. As a result it was highly maneuverable and an excellent performer on rough terrain.*

Thomas B. Jeffrey had been a long-established bicycle manufacturer in Chicago before he moved to Kenosha, Wisconsin, and opened an automotive plant. There Jeffrey soon became famous for his Rambler automobiles (named after his bicycles) and start designing trucks. (Jeffrey's first designs also included a pickup truck, which was the first to be equipped with pneumatic tires, but would be less famous than the Quad.)

Unfortunately, Jeffrey did not live to see his design roll off the production line. He died in 1910, leaving Charles T. Jeffrey to see the end results of his work, when the first Jeffrey Quad was finally produced in 1911.

Four-wheel drive

What was special about the Nash Quad was that it had four-wheel drive, which would make it a favorite for military use both at home and abroad, where good traction on poor and sometimes even nonexistent roads was an advantage.

The Nash Quad got its first taste of war during Brigadier General John J. Pershing's punitive expedition to New Mexico in 1916. (President Woodrow Wilson sent the expedition after Mexican bandit Doroteo (Pancho Villa) Arango attacked the town of Columbus.) The general didn't succeed in capturing Pancho Villa as planned, but he did discover the value of trucks for military purposes. When the expedition began, the

U.S. Army had just 1,000 motorized trucks in total. But before the 11-month campaign was over, many more were on order. Jeffrey Quads were among the 128 makes and models of trucks that eventually lined the border between the U.S. and Mexico. And the Quad's four-wheel drive made it extremely valuable as a military vehicle.

Nash buys Jeffrey company

The Quad would also catch the attention of Charles W. Nash (who had worked for the Durant-Dort Carriage Company). He bought the Jeffrey company in 1916. The name was changed to Nash, but one of the Jeffrey Company's chief assets—the Quad—would survive.

Four-wheel drive wasn't all the Quad had going for it. Nash Motor Company engineers gradually made additions to the original design that increased the Quad's functionality—the key selling point of a truck both in war and peace—as well as make further use of the truck's motor. Automatic dump bodies and powerful winches brought new applications, as well as road-grading gear, specially designed tank bodies, and trucks mounted with snowplows and drilling equipment. Even designs for ready-mixed concrete carriers were already being advanced as early as 1916, not to mention the improvements made to the actual truck chassis itself—including automatic locking differentials (which prevented the wheels from spinning and thus saved fuel and tires), internal gear drive axles (which delivered more power at lower engine speeds), and electric lights for night work as well as an electric starter. Quads and other trucks made by Nash would also soon have a load capacity of up to 2 tons (2.03 tonnes).

U.S. market leader

By 1918 the Nash Quad was the market leader in the United States; this was in addition to the massive truck production Nash as well as other companies were undertaking to supply the war overseas. Nash Quad production exceeded 11,000. The number didn't include the other two-wheel drive 1- and 2-ton (1.016- and 2.03-tonne) trucks, which were also being made by Nash.

Specifications

Country: USA
Year manufactured from: 1919
Engine: gasoline powered
Transmission: four-speed
Payload: up to two tons (2.03 tonnes)
Applications: delivery, military
Special features: four-wheel drive

Below: *The Nash Quad's ability to navigate uneven terrain made it a popular choice for wartime use.*

SCANIA

1919 Sweden

SCANIA-VABIS STANDARD TRUCK

By 1919, Scania-Vabis had decided to stop producing passenger cars and focus on trucks. By that time, the company was also using more fuel-efficient engines and had developed a four-speed gearbox.

Above: *Despite Sweden's cold climate, Scania Vabis trucks of the World War I era had open cabs.*

In 1911 Scania merged with Vabis of Sodertalje. While Scania was building its first truck at its plant in Malmo, in the province of Skane, Vabis (also in Skane) was test-driving a 9hp (6.7kw) engine along the cobbled city streets and rural roads of its hometown. But series production of trucks had only really taken off for both Scania and Vabis in 1907, when both companies launched trucks with tipping platforms.

The 1920s witnessed a steady series of minor improvements to the driver's working environment and ergonomics as well. Pneumatic tires made travel more comfortable and less noisy. Enclosed cabs kept out the worst of the weather and ventilation hatches enabled drivers to regulate airflow.

Improvements

However, heating systems for the cab had not yet been developed. Drivers were obliged to come up with their own ways to keep warm.

Still, there were other improvements. In 1922, the company patented a new type of carburettor. And by 1926, Scania-Vabis launched a new engine that could be adapted to different fuels. The Hesselman engine could run on fuel oil as well as gasoline. Unlike other carburettor engines, the Hesselman had an injection pump that distributed fuel to each cylinder via pressure pipes.

Specifications

Country: Sweden
Year manufactured from: 1919
Engine: gasoline powered
Transmission: four-speed
Payload: not available
Applications: multiple
Special features: wheels had ball bearings

OLDSMOBILE
1919 USA

OLDSMOBILE ECONOMY TRUCK

Above: *Competitive pricing and economical performance were important to the success of the Oldsmobile 1-tonner of 1919.*

The Olds Motor Works, the oldest U.S. motor vehicle company still in business, was founded by Ransom E. Olds in 1887 in Lansing, Michigan. He built its first vehicle, a three-wheel steam wagon, in the same year.

Right: *The vehicle could be supplied as a bare chassis/scuttle (hood, dashboard, etc) for mounting with a variety of special bodywork.*

Specifications

Country: USA
Year manufactured from: 1919
Engine: gasoline powered valve in head four cylinder
Transmission: three-speed selective sliding, with clone clutch
Payload: one ton (1.016 tonne)
Applications: light delivery
Special features: available with wooden or steel cab

In 1908 William Crapo Durant bought the Olds company. In 1919, Olds brought out its Oldsmobile Economy Truck—a 1-ton (1.016- tonne) commercial chassis with a deep channel section frame, which could be used as a bus, or as a light truck. A valve-in-head 40hp (29.8kw) four-cylinder engine powered it. It had a carburettor, three-speed selective sliding transmission, a clone clutch, a steering gear, and full electrics (starting, ignition, and lights). The wheelbase was 128in (3.25m) (the wheels were set out from the frame) and the truck rode on spoked wheels with solid rubber Goodyear tires. The truck's top speed was 22mph (35.4km/h). It was available with a wooden or steel cab with a solid roof and it had left-hand steering.

Available with or without a body

The chassis with an instrument board, windshield, cowl, and fenders were also available without a body, ready for mounting with stakes, a rack, a grain dump, or a box body. An "express" body was also available, with extra bracing and a tidier paint job for more money. It was a "conventional" design, with the driver's cab behind the engine.

The truck was discontinued in 1924, but it made quite a splash while it was around, largely thanks to its name—fuel was costly and getting costlier by the minute in 1919.

AUTOCAR

1920 USA

AUTOCAR

By 1911, Autocar decided to stop automobile production entirely in order to focus on the manufacture of commercial vehicles. In World War I, the company would produce two "conventional" models for military use.

Above: *Up to the mid-1920s Autocar concentrated on forward-control or "COE" trucks but conventional models like the one shown here eventually became available.*

In 1919, the company would bring out a line of cab-over-engine (COE) models of 2 to 5 tons (2.03 to 5.08 tonnes) payload. They were widely touted by the company as being easier to handle in traffic, as well as more economical because of their shorter wheelbases. Autocar would immediately embark on a coast-to-coast promotional trek. The 2- and 3-tonners that made the trip had 25.6hp (19kw) engines, while the four- and five-tonners were powered by 28.9hp (21.5kw) engines. (All their engines had four cylinders and were gasoline-driven.) 1919 also saw the company bring out a line of electric trucks, ranging in payload capacity from 1 to 3 tons (1.016 to 3.04 tonnes). By 1926, the company was producing 20 different models in an array of variations. Autocar had even returned to producing "conventional" cab-in-front engine models, such as the Type C, with a four-cylinder engine and same rear axle as the COEs.

Model F

In 1926, the lightest Autocar truck was the Model F, with a 97in (2.4m) wheelbase. The Model G had a 120in (3.04m) wheelbase and was available with an Autocar two cylinder engine in 1-, 1¼-, 1½-, and 2-ton (1.016-, 1.27, 1.52-, and 2.03-tonne) versions. The 2-ton (2.03-tonne) Models H and K were powered by 25.6hp (19kw) four-cylinder engines. Heavier models used 28.9hp (21.5kw) motors and solid rubber tires, their clutches, transmissions, and axles were made by Autocar.

Specifications

Country: USA
Year manufactured from: 1920
Engine: four cylinder gasoline powered
Transmission: chain drive
Payload: five tons (up to 5.08 tonnes)
Applications: multiple
Special features: shorter wheelbases therefore more economical in terms of fuel consumption

PIERCE ARROW TRUCK

🛠 **1920 USA**

PIERCE ARROW

Pierce Arrow trucks built nearly 7,000 of its cab-behind-engine trucks for export in 1918 alone. Their original truck had been a cab-over-engine design, but it didn't do well during testing so was abandoned.

Above: *Pierce Arrow of Buffalo, New York, were one of the earliest truck builders to introduce worm-drive axles when most other heavy trucks featured chain drive.*

Specifications

Country: USA
Year manufactured from: 1920
Engine: dual valve gasoline powered
Transmission: worm drive
Payload: not available
Applications: multiple
Special features: dual ignition, shockproof windshield glass and backup lights

The massive demand for Pierce Arrows overseas would make it one of the most respected truckmakers in North America at that time, ranked alongside Packard and Alco in terms of quality and expense. The period following the war was bad for truckmakers in that demand was way down, but good in that the U.S. government had finally realized the importance of building roads.

The United States Army had made its first transcontinental road trip from Washington, D.C., to San Francisco in 1919, in part to encourage the Government to invest in a national interstate road system for defense purposes. Dwight D. Eisenhower, a Captain in the Army at that time, went along for the ride. After 62 days of treacherous travel on dirt roads and crumbling bridges, Eisenhower and his contemporaries were convinced that better roads were the answer. The government would subsequently shell out $75 million on building roads and making improvements.

Fifth wheel standardized

Truck manufacturers were ready with improvements of their own. The fifth wheel was patented and became standardized on trucks in 1920. Backup lights were standard by 1922, and shockproof glass was used in windshields by 1926. But Pierce Arrow was ahead of the pack as usual, as their 1920 truck already had all of these features. It also had dual ignition and a dual-valved engine for more power and better fuel economy. And it could also be hooked up to different trailers.

PACKARD

1920 USA

PACKARD SIX-TONNER

By the 1920s, Packard trucks came in a variety of shapes and sizes—from delivery vehicles (also known as "hacks") up to trucks with an impressive 6 tons (6.09 tonnes) of payload capacity.

Specifications

Country: USA.

Year manufactured from: 1920

Engine: gasoline powered

Transmission: rear chain drive

Payload: six tons (6.1 tonnes)

Applications: multiple

Special features: could be customized with multiple bodies

Above: *Packard built a range of trucks with payload ratings from 1.5 to 7.1 tons (1.52 to 7.1 tonnes) until production ceased in 1923.*

The Packard brothers started building trucks under the Packard name in Detroit in 1905. Packard started the trend for "conventional" designs before World War I, when most truckmakers were using the cab-over-engine design. Packard would be the first of them to put the engine in front of the driver, emphasizing driver comfort and safety over load space and easy engine access.

Packard trucks were appearing with left-hand steering and chain drive on the rear axle as standard by 1915. Most also had solid rubber tires on steel spoke wheels and roll-down canvas windows (glass was also available). They came as tankers and with pups, with stake bodies and with flatbeds. The dump trucks had a cable system to lift the truckbed.

Focus on maintenance

Packard had been founded with good maintenance in mind. The story goes that brothers James and William Packard started what would originally be the Ohio Automobile Company in 1899, after a disappointing experience with their Winton automobile. The brothers had ordered a Winton car, only to have it towed to their house on delivery because it had broken down. Winton refused to give the brothers a refund or even repair their car for free, so they decided to build their own.

By 1911, Packard would own 7 percent of the U.S. truck market. This was in spite of how expensive the company's trucks were. In 1918, for example, Packard would sell a 3-tonner for more than $4,000.

VULCAN

1920s UK

VULCAN VSC

From 1918 to 1939, the small Vulcan company built some modern trucks, including the VSC. It was manufactured at first in response to a military demand, but was still in production after World War I.

Above: *A rare restored VSC. This truck was fitted with rubber tires and a reliable shaft drive.*

Specifications

Country: UK
Year manufactured from: 1920
Engine: six cylinder gasoline engine
Transmission: not available
Payload: 30 cwt
Applications: stake body
Special features: open cab

In 1902, the Hampson Brothers and E. Hope built a small light car that was displayed at the Liverpool Cycle and Motor Show. After this first success, a year later they founded the Vulcan Motor Manufacturing & Engineering Company Ltd. of Southport.

Growing output

The fledgling company soon found its feet, and production began growing rapidly. In 1906, about 115 cars were manufactured. To consolidate the company's success, a new factory was erected and the name was changed to Vulcan Motor & Engineering Company Ltd. During World War I, Vulcan had to turn to war work, and it built De Havilland aircraft frames alongside munitions and military trucks. After World War I, the commercial vehicle market became Vulcan's main source of income. The company, however, had many financial problems and in 1936 the bus and truck manufacturers Tilling-Stevens acquired the rights to Vulcan's truck production. The range of trucks was enlarged to 2 ½-ton (2.54-tonne), 3-ton (3.04-tonne), 5-ton (5.08-tonne), and finally 6-ton (6.1-tonne) just before the onset of World War II.

HUG

1922 USA

HUG MODEL T

In 1921, Hug decided to build its own trucks to suit its business specifications. Rugged but reliable, the Hug truck range was used during the twenties and the thirties for the big American construction projects.

Left: *The Hug Model T was a "no frills" machine built along simple rugged lines and aimed at the construction industry.*

With its open cab and its simple design, the Model T was a rugged truck designed for heavy applications. It was fitted with a Buda 34hp (25kW) gasoline engine and a Warner three-speed transmission with a Clark spiral bevel axle, as well as two-wheel brakes. The top speed was about 45mph (72km/h), double the speed of many other trucks from the same period. The rubber tires, with dual rear tires, were uncommon for a severe-duty truck and despite a small size the Model T could carry a load of up 2 tons (2.03 tonnes). It was bodied with a dump or with a concrete mixer. In the middle of the twenties, three other models were built, with payloads from 1½ tons to 3½ tons (1.52 to 3.56 tonnes). Like all the Hug trucks until 1932, they were powered by Buda engines. Hug used subcontractor-built components like Clark, Columbia, Timken, or Wisconsin axles and Brown-Lipe or Fuller transmissions. These trucks kept the rugged, "no-frills" appearance, and in 1927 a massive six-tonner with a seven-speed gearbox appeared. After 1930, the range was completed with the "Roadbuilder" series and, for the first time, the series was available with Caterpillar diesel engines. At the beginning of the war, Hug manufactured some 6x6 military trucks, but no contracts were concluded and 1942 was the last production year for the company.

Specifications

Country: USA

Year manufactured from: 1922

Engine: four cylinder Buda engine, 34hp (25kw)

Transmission: three-speed

Payload: two tons (2.03 tonnes)

Applications: dump truck, concrete mixer

Special features: rubber tires, open cab

CHENARD AND WALCKER

Mid 1920s France

CHENARD & WALCKER

For most people, Chenard & Walcker was a car manufacturer. However, it was also an innovative truck builder. In France, its contribution to the semitrailer combination was significant.

Above: *The Chenard & Walcker UT was one of the first tractors ever built.*

Specifications

Country: France
Year manufactured from: 1927
Engine: four cylinder gasoline powered 20hp (15kW)
Transmission: three-speed
Payload: 1.96 to 9.84 tons (two to ten tonnes)
Applications: semitrailer or trailer hauling
Special features: short wheelbase truck, trailer coupling system

The prestigious Chenard & Walcker company was founded in France in 1901 and was famous for its cars. In 1920, the company was the fourth-largest car manufacturer in the country.

FAR company

At the end of the World War I, the company created a new department, the FAR company, in order to exploit the patented Lagache et Glazsmann trailer-coupling system. Their trailers and semitrailers initially had a very limited distribution, and it was not until the sixties that they spread throughout Europe. Chenard & Walcker wanted to prove that a small short-wheelbase tractor could haul more payload than a rigid truck on short distances. The tractor could haul a trailer to a warehouse, uncouple the trailer while it was unloaded, and return to base for a further haul. This system was more profitable than using a rigid truck. The company built two- or three-axle bonneted tractors. Some heavy 6x4 axle configurations, with chain drive, were designed for logging applications or special big haulages. Some of them were sold as omnibus tractor units in Holland.

MAN

1924 Germany

MAN DIESEL TRUCK

MAN Nutzfahrzeuge AG started producing commercial vehicles for civilian and military use in 1915. From then on MAN military trucks formed the backbone of the German army truck fleet.

Specifications

Country: Germany

Year manufactured from: 1924

Engine: diesel powered, direct injected

Transmission: worm and chain drive

Payload: four tons (four tonnes)

Applications: multiple

Special features: first direct-injected diesel engine

Above: *MAN pioneered the modern direct-injection diesel with this 4-tonner, exhibited at the 1924 Berlin Show. It was powered by a 4-cylinder, four-stroke engine, developed in 1924.*

The German truckmaker had a long history of innovation. Back in the late 1800s, the company had manufactured steam-powered commercial vehicles, and had experimented with the diesel engine in the very early stages of its initial development by Rudolf Diesel. In fact, the company worked with Diesel to build the world's first diesel engines at two plants in the cities, which appeared in the company's full name: Maschinenfabrik Augsburg-Nürnberg. But the first diesel engines would be stationary, and MAN's first commercial vehicle, introduced in 1915, would actually have a gasoline engine.

The company's first truck would be produced under the Saurer license. It was a four-tonner with a 36hp (26.8kw) engine and chain drive. Still, MAN would eventually bring out a diesel-powered truck—a 45hp (33.5kw) model with the world's first direct-injection diesel engine, introduced in 1924. In an indirect-injection diesel engine fuel is injected into a small pre-chamber, which is connected to the cylinder by a narrow opening. The initial combustion occurs in the prechamber. This has the effect of slowing down the rate of combustion. In a direct-injection diesel engine, fuel is injected directly into the cylinder. The piston incorporates a depression, which is where the initial combustion takes place. Direct-injection diesel engines are generally more efficient than indirect-injection engines.

AMO-F-15
1924-31 USSR

AMO-F-15

Russian automobile and truck manufacture had developed late compared to the rest of the world, and the Soviet Union was afraid of being left behind, especially in time of war.

Right: *The AMO F-15, one of the Soviet Union's first home-built trucks, was based on a Fiat design. It was powered by a four-cylinder gasoline engine.*

Specifications

Country: Russia/USSR
Year manufactured from: 1924
Engine: four cylinder gasoline powered
Transmission: four-speeds
Payload: two tons (2.03 tonnes)
Applications: multiple
Special features: chassis would also be used for fire trucks, buses, and armored cars

The Russian company Zil had started manufacturing cars in 1916. By 1924, the company had begun making the AMO-F-15 under the AMO name. It had a four-cylinder, 2,637.6ci (4,396cc) engine, and could reach speeds of up to 35 to 50mph (56.3 to 80.4km/h) (depending on terrain and load). It weighed in at 4,233.6lb (1,920kg), had a four-speed gearbox, a 10ft (3.07m) wheelbase, and road clearance of 8.8in (22.5cm).

The first body fitted to the AMO in 1924 was a cargo type. But the chassis, in true Russian style, was also used for a multitude of other purposes—in 1925 it was fitted out as an ambulance, in 1926 as a 14-passenger bus, and in 1927 as a fire truck, a staff car, and an armored car, the BA-27.

Help from the U.S.

The formerly czarist AMO plant was reorganized and expanded in 1929 thanks to a technical assistance agreement with the Arthur J. Brandt company of Detroit. (In 1917, the plant had been equipped with U.S. machinery.) Design work for the expansion was handled in Brandt's Detroit office and plant and American engineers were sent to Russia. The Soviets also signed an agreement with the Ford Motor Company in May 1929 to purchase $13 million worth of parts and automobiles in exchange for technical help.

GMC

1925 USA

GMC K SERIES

William Crapo Durant originally formed the General Motors (Truck) Company in 1911 to handle sales for several truck and passenger vehicle-manufacturing companies that Durant had acquired.

Specifications

Country: USA

Year manufactured from: 1925

Engine: four cylinder, gasoline powered

Transmission: dual rear chain drive

Payload: five tons (up to 5.08 tonnes)

Applications: multiple

Special features: chassis came with or without body

Above: *The GMC K41 model in articulated form could take a 10-ton (10.1-tonne) payload and featured shaft drive and solid tires.*

Therefore when the first GMC trucks were produced in 1912, they were an amalgam of the trucks made by other builders. Certain components would eventually be standardized. By 1913, GMC trucks made the switch to a conventional four-cylinder L-head Continental engine. But other characteristics of the newly named GMC Rapid and GMC Reliance trucks remained the same. For example, GMC's Rapid truck continued to have a four-speed transmission and was rated up to a capacity of 2 tons (2.03 tonnes), while GMC Reliance trucks retained forward control and three-speed transmission and it was offered in payload capacities up to 5 tons (5.08 tonnes). Both models also retained a double chain drive.

New technology

Meanwhile, GMC soon began to experiment with its own technology, and brought out an ill-fated range of electric trucks in 1913. The truck batteries had to be charged every 40 miles (64km) and they were eventually discontinued.

By 1925, GMC brought out its gasoline-driven K series, ranging from one-tonners to highway tractors. The longest available wheelbase—187in (4.74m)—was available on one three-tonner and one five-tonner—the K-71B and the K-101B respectively. The latter's engine could provide up to 32.4hp (24.1kw) of power. Timken made front and rear axles on all but the lightest K models. The lightest model had GMC axles and pneumatic tires as standard equipment.

WHITE
🛠 **1926 USA**

WHITE MODEL 56

The White Motor Company brought out its 2-ton Model 56 in 1926. White trucks had come a long way from the days when drivers had to hand-crank them to start them up. By 1926, the trucks had full electrics.

Right: *White trucks of the mid-1920s had an advanced design. Chain drive had been abandoned for shaft drive, as featured on this vehicle.*

Specifications

Country: USA
Year manufactured from: 1926
Engine: side valve with four cylinders, gasoline powered
Transmission: double reduction two piece shaft driven, four-speeds forward
Payload: two tons (2.03 tonnes)
Applications: multiple
Special features: used a centrifugal water pump for cooling.

The company had stopped producing passenger cars in 1918 because they wanted to focus solely on truck and bus manufacturing. By 1919, the standard chain drives on White heavy trucks were obsolete. Heavy chassis 3 to 5 tons (3.04 to 5.08-tonnes) were equipped with double-reduction two-piece shaft drives instead. Brakes had also seen significant improvements. And cast vertical tube-type radiators with removable heads were adopted, as were removable cylinder heads. Removable heads would make repairs easier and more economical for White truck owners.

Postwar sales dip

After the war, White truck and bus sales dipped, but would recover, thanks in part to the introduction of the Model 50 in 1921. White continued to build trucks with 5-ton (5.08-tonne) capacity through the 1920s, including the Model 45-D Power-Dumping Truck, with a 28.9hp (21.5kw) engine, introduced in 1922. By that time, both dual and single rear wheels were available with pneumatic tires, at extra cost of course.

In 1925, the White company offered a wide range of truck and bus chassis. The ¾-ton (0.76 tonne) Model 15 had a 22.5hp (16.7kw) four-cylinder engine, as did the Model 20 two-tonner. The Model 45, for its part, had a 50hp (37.2kw) engine. By 1926, engine power on some six-cylinder models (such as the Model 54 bus, with overhead valve engine) was enabling top speeds up to 100mph (160km/h). By this time airbrakes were also standard fitment. A side-valve, single-cast monobloc engine with four cylinders powered White's two-ton Model 56. The Model 56 had four-speed transmission going forward, and used a centrifugal water pump for cooling.

VOLVO

1928 Sweden

VOLVO SERIES 1 TRUCK

Volvo started producing cars in 1927. But as early as December 1926, a design for the Swedish company's first medium-duty truck (Series 1) was in the works. It was to be a very successful vehicle.

Above: *Compared to some manufacturers Volvo were latecomers to the truck market but, from the outset, they set extremely high standards of engineering and comfort.*

The first Series 1 rolled off the production line in Gothenburg, Sweden, in February 1928. Even though it was the first truck made by Volvo, it held its own when it came to driver comfort and new technology.

The very first Series 1 truck was shaft driven, with pneumatic tires. This configuration was unusual at a time when many trucks were commonly fitted with chain-drive systems and solid rubber tires. Furthermore, while many truck drivers were still forced to sit on a bare wooden seat out in the open air completely exposed to the elements, the Volvo Series 1 could be fitted with a cab on demand. Sweden's cold climate probably had much to do with the latter. But heating was still an issue. In this truck, like in other trucks up to the mid-1930s, hot air was provided by the engine, which entered the cab through the metal sheet that separated the engine from the cab.

The engine, designed by Volvo, had four cylinders and was gasoline-powered. It could only reach speeds of up to 31mph (50km/h), depending on payload and terrain, and it could only deliver 28bhp (20.8kw), but the Series 1 did have a three-speed gearbox—also designed by the company. It was also reportedly stronger than it seemed—while the Series 1's official payload was limited to 3,307lb (1,500kg), customers frequently boasted they could use it to carry loads up to double the official figure.

Optional cab

And you didn't have to buy the cab—the Series 1 truck was also available as a chassis. A standard cab was produced by an independent body manufacturer (the same one that produced the Volvo car bodies), but it could be ordered and fitted onto the truck in the Volvo factory prior to delivery. (Many customers did opt to take delivery of the Series 1 without the standard cab, and then order a cab to their own specifications; a testimony to the vehicle's multiple applications at that time.)

The Series 1 proved to be much more successful than anticipated, selling more quickly than the cars Volvo had so far produced. The first 500 trucks sold out in six months, much sooner than expected. (Volvo had expected the first batch to last for up to two years.) Volvo made a second 500-truck batch, but due to the unexpected demand, the company didn't have time to make many improvements. The company would produce another batch of 500 Series 1 four-cylinder trucks nearly identical to the first trucks, with engines that were no more powerful than the originals. Volvo did, however, manage to reduce the number of axles on the second batch. One of the Series 1's previous two rear axle ratios (the slower one) was eliminated, resulting in a slower top speed but leading to improved drive and handling.

The company would take the opportunity to make several improvements to its original truck when it launched Series 2.

Change of track dimension

Foremost among them was the decision to change the track dimension of the truck. The Series 1 had had the same narrow track dimension as the first Volvo car—only 4ft 3in (1.3m)—a poor design for the roads of that day, which often consisted of nothing more than a pair of wheel ruts left by horse-drawn carts. The latter were usually about 4ft 11in (1.5m) wide, and the Series 1 truck wheels were too close together to fit. Volvo responded to this by moving the rear springs to the outside of the frame rails, thus giving the Series 2 a wider track of 4ft 9in (1.46m). (The Series 1 truck springs had been attached directly underneath the frame rails.)

The Series 2 truck was also less expensive for Volvo to make. The Series 1's wooden steering wheel was, for example, exchanged for a more standardized Bakelite steering wheel. Volvo's Series 2 truck matched its predecessor's success. It also found favor with bus companies, which used the chassis as the base for a wooden-framed bus body covered by sheet steel.

Specifications

Country: Sweden

Year manufactured from: 1928

Engine: gasoline powered, four cylinders

Transmission: shaft driven, three speeds

Payload: up to 6,614lb (3,000kg)

Applications: multiple, body made to order

Special features: closed cab option

Below: *Volvo built high margins of strength into their original Series 1 truck because it was common for users to double the recommended payload at that time.*

VOLVO
1928 Sweden

VOLVO SERIES 2

Working as engineers making bearings in the Swedish company SKF, Assar Gabrielsson and Gustaf Larson dreamed of building a car suited to the Scandinavian climate. That's why they founded Volvo in 1924.

Above: *Old fashioned but reliable and tough, the Volvo series 1 and 2, designed for the local market, were the first commercial successes of the Swedish company at the end of the twenties.*

The name "Volvo," which means "I roll" in Latin, was a reference to the bearings in action, but it was not before 1928 that the first truck appeared. In 1927, the first Volvo car was launched. It was a real challenge, because Gabrielsson and Larson wanted to build a complete car, from the frame to the engine. The company had planned to manufacture cabriolets (OVs) and sedans (PV4s), but despite lots of improvements to its vehicles it was not in profit because of the poor local demand. Some chassis were modified as light trucks and the first ten car prototypes were sold as platform trucks by cutting into section the back of the frame. Some customers acquired the old-fashioned OV or PV4 and converted them into light trucks. The real first truck was shown in 1927.

Basic model

The first Volvo truck was a 1.47-ton (1.5-tonne) payload capacity vehicle, but it could carry up to double that weight despite a modest 28hp (21kw) four-cylinder engine. Two wheelbases and two different axle ratios were available. The truck was a great success: 497 were sold by 1929. Volvo planned to sell the model over two years, but after only six months it was sold out. The gearbox and the shaft drive were extremely robust, as was the Swedish steel frame. Most of the trucks were sold as chassis-cab, but Volvo was able to fit an enclosed cab built by an independent coachbuilder, the Atvidaberg Company. At this time, the other imported trucks were not designed for the freezing Swedish temperatures and they couldn't stand up to the climate. Volvo didn't expect such a success and had to launch a new series to face the demand as soon as possible. The series 2 was fitted with the same underpowered engine, but the rear axle ratio was reduced to a single one in order to improve the reliability. The top speed was slower, but the truck was tougher when carrying payload. The main difference between the series 1 and the series 2 was the track dimension. In fact, the series 1 kept the car's track (4ft 4in/1.30m) and this was a problem when driving on the so-called roads (mainly smudged tracks). The truck had poor stability and couldn't follow the horse-drawn cart tracks, which were wider. Hence the tracks were widened to 7ft (1.46m) by moving the rear springs to the outside of the frame rails. Like the series 1, the series 2 was a simple, rugged, and reliable truck. With the introduction of the new six-cylinder LV 60 series in 1929, the older series received a new tradename. The series 1 became LV 40, LV 41, LV 42, or LV 43 depending on the wheelbases and the rear axle ratio as seen before, and the series 2 was called LV 44 or LV 45 (two wheelbases). The series 2 was quite successful but it took more than one year to move all the 500 trucks in stock. The demand remained high, but the six-cylinder LV 60 was now in competition. For the company, the truck business was very profitable and the Volvo name found favor with bus companies and truck operators.

Specifications

Country: Sweden

Year manufactured from: 1928

Engine: four cylinder gasoline powered, 28hp (21 kW)

Transmission: three-speed selective sliding

Payload: 1.47 tons (1.5 tonnes)

Applications: stake body, parcel delivery, dropside platform

Special features: shaft-drive axle, four cylinder flathead engine, wider track than the series 1

Below: *In comparison with the series 1, the series 2 was remarkable for its wider axle front and rear tracks designed to cope with the rugged roads in Sweden.*

INTERNATIONAL HARVESTER
1928 USA

INTERNATIONAL HARVESTER SIX-SPEED SPECIAL

By 1928, the International Harvester Company (IHC) of America had already established a name for its trucks, and not just back in rural America, but also for multiple applications in cities.

International Harvester trucks were being sold in countries as far away as Japan, where IHC chassis were marketed for multiple applications, including commercial hauling, municipal work and passenger transportation. Commercial haulers were available in sizes from 3,000 to 10,000lb (1,360.5 to 4,535kg) payload capacity, with Speed Truck light- to medium-duty vehicles in 1¼-, 1½-, 2-, and 3-ton (1.27-, 1.52-, 2.03, and 3.04-tonne) versions and heavy-duty trucks in 2½-, 3½-, and 5-ton (2.54-, 3.55-, and 5.08-tonne) models. They could be fitted with express bodies, panel bodies, dump bodies with an under-body hoist, and special bodies for any purpose. For municipal work, they could be fitted as fire trucks, garbage trucks, ambulances, police wagons, street maintenance vehicles, etc. For passenger transportation the sky was the limit.

Above: *Launched in 1928, International's "Six Speed Special" one-tonner featured a two-speed rear axle, which, combined with the normal three-speed gearbox, provided six forward and two reverse ratios.*

Six-speeder returns to roots

The International Harvester Six-Speed Special truck, introduced in 1928, was the company's first truck with a two-speed rear axle (for going in reverse—the truck also had six speeds going forward.) It was specially designed for off-road rural haulage and became popular immediately. (In fact, earlier versions of the Six-Speed Special had already contributed to doubling International Harvester truck sales in the last three years.)

The Six-Speed Special owed at least some of its popularity to the fact that it was a return to the company's farming roots (IHC had started out producing one of North American's first motorized reapers.) But the Six-Speed Special double reverse wasn't the sole contributor to the company's success.

Marketing goldmine

In 1928, IHC management hit on an idea that would become a tried-and-true truck-marketing goldmine thereafter. In the age of America's love affair with the romance of the desert (Howard Carter had discovered Tutankhamun's tomb in 1922) why not promote a truck used by someone who had conquered it? The idea soon paid off, in the form of an article/advertisement in the pages of the October 1928 edition of *Scientific American*.

The article, with the hero/explorer in the foreground and an International half-ton panel truck in the back, told the story of Sir Charles Markham and Baron Frederick Von Blixen-Fineke's 3,000-mile (4,828km) trek through the Sahara desert in the world's first four-wheel truck to do so. The truck had a fully covered cab and body, with unglazed windows on all sides to allow the desert air to circulate more freely (though as one would expect there was a system for covering the windows in case of sandstorms). It had an engine in front, and remountable pneumatic tires (as well as several spares fastened to the body) and appeared to have been tough enough to endure the extreme temperatures and sand of the world's biggest desert.

The article describes in detail a first-person account of the journey made by Sir Charles, but of course it does not go so far as to give any details as to whether the truck suffered from any unforeseen mechanical problems on the journey or not.

Specifications

Country: USA
Year manufactured from: 1928
Engine: gasoline powered
Transmission: six-speeds forward, two speeds reverse
Payload: up to 5.08 tonnes (five tons)
Applications: multiple
Special features: specially designed for off road haulage

Left: *With its roots in the agricultural industry, International Harvester produced tough trucks that were highly popular among farmers who often needed to traverse unmade roads.*

GRAHAM BROTHERS
1928 USA

GRAHAM BROTHERS SIX-CYLINDER G-BOY

As 1928 drew to a close the Dodge Brothers' and the Graham Brothers' truck companies had already merged. While the Dodge name lives on to this day, it was Graham Brothers that saved Dodge from extinction.

Specifications

Country: USA
Year manufactured from: 1928
Engine: six cylinder, gasoline powered
Transmission: four-speeds forward
Payload: 1¼ tons (1.27 tonnes)
Applications: multiple
Special features: four wheel brakes

Above: *Graham Brothers trucks were built around Dodge running units like this 2.5-tonner which featured a six-cylinder engine and spiral bevel rear axle.*

The Graham Brothers Truck Company had established a name for itself as early as 1912, when a Graham truck participated in a United States Army road test conducted to determine whether or not trucks could replace horses, mules, and/or trains under field conditions. The Graham was among nine trucks selected to supply a provisional regiment during a practice march from Dubuque, Iowa, to Sparta, Wisconsin. The Graham truck was one of three trucks used to replace the railroads in resupplying the field trucks out of the two main supply bases in Dubuque and Madison.

Indeed, the Dodge name could have disappeared without a trace in 1924 when the company first merged with the Graham Brothers. Both founding Dodge brothers, John and Horace Dodge, had died in 1920, at which time the commercial vehicle chassis they produced accounted for about 10 per cent of the Dodge company's total volume. But by 1925, Dodge's chassis production had doubled largely because of its association with the Graham Brothers. The company, which originated in Evansville, Indiana but later moved to Detroit, Michigan, built truck bodies on the Dodge commercial chassis. By then, the trucks had come to be sold under the Graham Brothers name.

Engines built by Dodge

By 1928, Graham Brothers Trucks were available with four- or six-cylinder engines built by Dodge. They also had four-wheel brakes on types of vehicles, including a light delivery van with a 110in (2.79m) wheelbase, a 120in (3.04m) wheelbase commercial truck, and trucks with load capacities from 1.25 tons (1.27 tonnes) (the G-Boy) to 2.5 tons (2.54 tonnes). Wheelbases for these ranged from 130in to 165in (3.3m to 4.19m), and all also had four-speed transmission.

DELAHAYE

1929 France

DELAHAYE PS 92

Delayahe was one of Europe's most advanced car manufacturers at the beginning of the 20th century. The company was founded in Tours, but quickly re-established itself in Paris under Emile Delahaye.

Above: *The Delahaye PS 92 was one of the early fire trucks produced in France.*

Specifications

Country: France
Year manufactured from: 1929
Engine: four cylinders overhead valves engine, 30 kw (40hp)
Transmission: not available
Payload: not applicable
Applications: fire engine
Special features: engine driven pump

The company was famous for fire engines and fire apparatus. Hence the major part of Delayahe's income was the truck business, especially during World War I. Delahaye became renowned for its beautiful engines: all the Delayahe engines were fast and powerful—exactly what a fire truck needed when a rapid response was required.

PS 92

After World War I, Delayahe brought the fire engine and fire truck ranges up to date. The best example of this modernization is the PS 92. The Premier Secours Model 92 ("First Aid" fire truck) was very popular in all French rural districts until the fifties. This fire engine was smaller than previous models, but had an improved practicality. In many ways it was a thoroughly modern fire truck, with a pump driven by the engine (called "Petit cheval", i.e. "Little Horse"), a 3182-liter (700-gallon) water supply, an independent hose reel and a six-man crew. It had a 30kw (40hp) four-cylinder engine, but the radiator size was increased to avoid overheating when the pump was being operated.

REO SPEED WAGON

🛠 **1929 USA**

REO SPEED WAGON STANDARD 1½-TONNER

The REO Speed Wagon was named after its designer, Ransom Eli Olds (1864–1950), who founded the Olds Motor Vehicle Company in Lansing, Michigan, in 1897 and later the REO Motor Company.

Above: *Reo first adopted the "Speed Wagon" name for its trucks in 1915 and progressively updated the range through to 1939 when the last Speed Wagons were built. This is a late 1930s two/three-tonner with a tipping body.*

One of his most popular vehicles, the REO Speed Wagon (from which a popular rock group of the 1970s later took its name) was originally fitted out as flatbed pickup truck. It was very high speed and heavy duty and was considered a milestone of transportation history. Its rugged ability to withstand the rough roads of North America in the early 1900s meant it would eventually be developed to service multiple applications. Over the years it would see service as a tow truck, fire truck, fuel delivery truck, in short, for all applications that required a rugged workhorse design.

Advertisements for the truck in the 1920s boasted, "This REO Speed Wagon fits every business... Even in those lines where you would naturally think only in terms of 10-ton trucks, we find Speed Wagons rendering splendid—and owners tell—more economical..."

REOs (cars included) were popular all the way up to the time of the Great Depression, and were reportedly capable of logging 100,000 miles (160,930km) without abnormal wear and tear. Fans claim the

Speed Wagon set the standard that most companies producing trucks at that time tried to imitate.

The first REO Speed Wagon was a ¾-tonner. It was advertised as being capable of up to 22mph (35.4km/h)—although those who used it claimed it could reach 40mph (64.3km/h)—even when fully loaded. The Speed Wagon had an electric starter and a 45hp (33.5kw), four-cylinder engine. It also had a 13-plate clutch and three-speed selective sliding gear transmission, and bevel drive.

Several models available

By the end of the "Roaring Twenties" the Speed Wagon chassis was available in several models, including the Junior ½-ton, the Tonner, the Standard 1½-ton with two choices of wheelbase, the Master (2 tons), and the Heavy Duty (3 tons).

The 1929 Speed Wagon had a six-cylinder engine, with L-type head, 3.25in (8.2cm) bore, and 4in (10.1cm) stroke. It had four-point suspension, seven-bearing crankshaft, and aluminum-alloy pistons. It also had a thermostatic temperature control, semi-automatic spark control, and a single-plate clutch. As was usual at that time, the chassis could be customized with a coupe cab (or not) and there were three standard body styles available (full panel, stake (wood panel flatbed) or Express, available with or without a canopy top.) The vehicle rode on spoked, malleable iron artillery wheels and pneumatic tires, with a suspension that included extra-long semi-elliptical springs. Speed Wagons of the time also had four wheel internal hydraulic brakes, a standard 23in (58cm) wheelbase, and an irreversible steering gear, not to mention an automatic chassis oiling system.

But more than that, Olds' Speed Wagon trucks were popular for their stylish designs, as well, not to mention the backing of the familiar Olds name. At its height in the 1920s, the factory in Lansing, Michigan employed about 6,000 people.

Customer testimonials

By 1930, testimonials alone would sell the trucks: "It might interest you to know that my second two-ton REO Speed Wagon has passed the 65,000-mile mark with a repair expense of but $201.10, and I expect to run it another 66,000 miles," reads a customer testimonial published in 1930. REO stopped making cars altogether in 1936 and changed hands several times starting in the 1950s, eventually merging with the Diamond T division of the Cleveland-based White Motor Company.

Specifications

Country: USA
Year manufactured from: 1929
Engine: four cylinder
Transmission: three-speed selective sliding
Payload: 3/4 ton (0.76 tonnes)
Applications: delivery
Special features: electric starter

Below: *The Speed Wagon was ruggedly built from the beginning and boasted powerful performance. A variety of body types for different applications was available.*

VOLVO

1929 **Sweden**

VOLVO LV SERIES

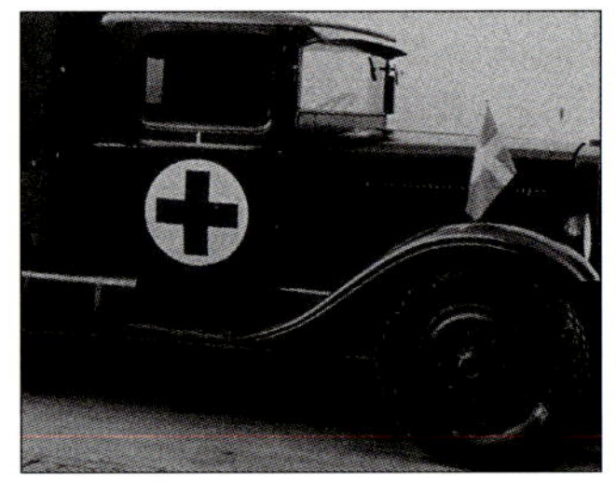

Volvo started producing the LV series in 1929. The very first models were light- to medium-duty with two-wheel brakes and wooden-spoke wheels. But Volvo wanted to bring out something more modern.

The LV71 and the LV73 were designed for different types of applications. The former was primarily intended for distribution on roads, while the latter had a larger margin for overload and was therefore better suited for off-road construction. The mechanical components of all four trucks were conventional. They had side-valve engines, four-speed gearboxes, and single-reduction rear axles.

But they weren't as powerful as Volvo's next models—the LV66 and LV68, which were available with more powerful overhead valve gasoline Hesselman engines. The LV75 came next, in 1933. Its design was a direct response to limits on the axle pressure allowed on the poor European roads of the day. Volvo engineers wanted to distribute the weight more evenly between the axles of the truck, so they moved the cab behind the engine and front axle. It had a 65bhp (48.4kw) side-valve gasoline engine. It was soon put to use as a chassis for light- to medium-duty buses in rural areas.

Next came the LV76–78s series, light-duty vehicles also intended for distribution applications. Their front ends resembled those of Volvo cars of the day, but their fenders were wider, to allow for bigger tires.

Suited for heavier use

In 1936, the LV79, was introduced, with heavier chassis components and double-mounted rear wheels, for heavier use. But there was still one thing missing in the Volvo truck product lineup—a really powerful engine, for example, a truck with enough engine power for snowplow applications. The LV29 series of trucks (otherwise known as the "Longnose" because of the truck's very long hood) with a gasoline/Hesselman was introduced in 1937.

Above, top: *The LV60 series was Volvo's first truck with a six-cylinder engine, seen here as an ambulance.*

Above: *The first LVs included the closed cabs and engine-before-cab elements, by now common to most truck designs.*

Specifications

Country: Sweden
Year manufactured from: 1929
Engine: side valve (later overhead valve) gasoline powered
Transmission: four-speeds, single reduction at rear
Payload: varied
Applications: multiple
Special features: engine forward design

SINGER
1929 UK

SINGER TWO-TON LORRY

Like many companies of its day, Singer started building bicycles and ended up building cars. Along the way, the company also tried its hand at producing commercial vehicles.

Above: *Singer first attempted commercial production in 1929, with the introduction of a 2-ton (2.03-tonne) truck.*

Specifications

Country: England
Year manfactured from: 1929
Engine: gasoline powered
Transmission: not available
Payload: two tons (2.03 tonnes)
Applications: delivery, multiple
Special features: chassis also used for bus

George Singer had been a bicyclemaker in Coventry, England, since 1874. He soon graduated to producing motorcycles and in 1904 produced his first car at a factory on Canterbury Street.

First commercial truck

Singer first attempted production of commercial vehicles in 1929, with the introduction of a 2-ton (2.03-tonne) truck and a 20-seat bus by the company's newly formed Industrial Motors division. But by 1932, the division was closed and the attempt at commercial production was abandoned, because of the company's increasingly shaky financial situation.

In 1929 Singer had been third in the list of British car production, after Austin and Morris, but by the mid-1930s its market share had slumped. It was slowly being squeezed out on its home turf by competitors, including successful American manufacturers like Ford.

By 1935 Singer had recorded a sunstantial loss in profits. (It hadn't helped that managing director William E. Bullock bought a building in 1927 that was not at all suitable for car production.)

THE FIRST VEHICLE
BUILT BY ERF LTD.
AND SOLD TO
W.F. GILBERT
LEIGHTON BUZZARD
1st SEPTEMBER 1933
ERF LTD.
SUN WORKS
SANDBACH
MJ 2711

chapter 4

War Returns

Material shortages and economic hardship characterized the period between the wars. The truck-manufacturing industry would be not be spared. Postwar flooding of the market with used military trucks would hit sales, while steel and fuel shortages made manufacturing and operating trucks more problematic as well.

Above: *The Volvo LV 80 series was a new trucking concept created to increase load capacity.*

Left: *The CI4 was the first truck to be produced by ERF in 1933 after the company founder, Edwin Richard Foden left his family's Foden factory to set up his new venture.*

Still, there was a need for trucks to rebuild postwar Europe and Asia, even if shortages persisted. The ability of truck manufacturers to survive would be completely dependent on their ability to somehow fill a need that was most definitely there.

In short, there were opportunities, no matter how small, to at least eke out a living during the economically challenging interwar period. Some of these opportunities could be exploited via technological development. For example, in those hard times, it was imperative that manufacturers made running a truck less expensive and easier. So makers started offering diesel engines instead of gasoline ones, and tried to lighten trucks so they would consume less fuel. Steel alloys were invented to replace metals that were too heavy or in short supply.

Maintenance was another area where truck-makers found they could make money. Some would be able to capitalize on their ability to manufacture and/or replace the parts of trucks they had manufactured for military use during the war and which were now flooding the market. Other manufacturers would survive by simply continuing to make sales no matter what—taking horses and wagons in place of cash and devising elaborate financing schemes.

BEDFORD

1930s UK

BEDFORD "W" SERIES

In Bedford nomenclature the first letter indicates model. Some historians of British trucks say a "W" means a chassis made to wartime specifications, although the "W" series reportedly came out in the early 1930s.

Above: *The English company Vauxhall Motors started producing the Bedford brand of trucks in the early 1930s. The first trucks of the "W" series were four-wheelers.*

Bedford was a subsidiary company of Vauxhall Motors, which was a well known provider of commercial vehicles for the United Kingdom and export markets. It began as the Vauxhall Iron Works, founded by Alexander Wilson in 1857. The company originally built machinery, such as pumps and engines, for the marine industry. It was named Vauxhall because it was first located on Wandsworth Road, in Vauxhall, London.

By the turn of the century Wilson had left the business and the remaining board directors became interested in starting production of a "horseless carriage." The company had already started producing a gasoline-powered launch (one of the first in marine engineering). In 1903, the company was manufacturing a 5hp (3.7kw) single-cylinder car, available to the public for 130 guineas.

The vehicle sold well, and the board of directors decided to expand the production of gasoline-powered vehicles. But in order to do so, the factory would have to move out of London. The company acquired a site in Luton, in Bedfordshire, where it moved in 1905. The company continued to do business under the name Vauxhall Iron Works until 1907, when the modern name of Vauxhall Motors Ltd. and the company's griffin logo was adopted.

Car production continued. Then, in November 1925 Vauxhall was acquired by General Motors. Chevrolet trucks had been on sale in Britain since 1923, but were costly owing to import tax. In 1925 GM set up an operation in Hendon, in Middlesex, to prepare imported vehicles and introduced Buick, Cadillac, Chevrolet, La Salle, Marquette, Oakland, Oldsmobile, and Pontiac cars, as well as Chevrolet and GMC trucks.

But with the country in the grip of a worldwide Depression, there was no little public hostility to imported products. (More than two million people were unemployed in the United Kingdom alone at this time.) So General Motors decided to camouflage its foreign roots by building a completely "British" truck using Vauxhall's Luton plant.

The British truck is born

By the early 1930s, assembly of the last Chevrolet and GMC trucks and vans in Britain had ended and the manufacture of Bedford trucks had begun. In 1930, the first true Bedford was on the road.

The first trucks of the "W" series were four-wheelers—a 30cwt truck and a 2-ton (2.03-tonne) version. The trucks were probably equipped with Bedford's own four- or six-cylinder gasoline-powered engines, which could produce up to 27hp (20.1kw).

The earliest "W" model trucks most probably had four-speed gearboxes attached to bevel drives on the rear axles. It is also safe to assume they had Dewandre servo brakes because those would also appear on the company's commercial passenger vehicles at that time. The trucks proved so successful that by 1937 sales of the Bedford "W"series had reached 30,000.

"W" series on hold

Historians report the "W" series was put on hold during World War II, when the production of civilian and commercial vehicles in Britain was suspended in favor of focusing the nation's efforts on the manufacture of military vehicles. Bedford trucks manufactured during World War II included the "MW" and "OY" types, with modified tractors. Bedford also produced the "QL" 4x4 for military use. After World War II, commercial and civilian production resumed, and Bedford resurrected many of the elements of the original "W" series.

Specifications

Country: UK
Year manufactured from: 1931
Engine: four or six cylinder gas powered
Transmission: four-speed, bevel axle
Payload: up to 2.03 tonnes (two tons)
Applications: delivery, military
Special features: servo brakes

Below: *Built in Luton, in England, by Vauxhall Motors, Bedford trucks dominated the UK medium-truck market in the 1930s. Vauxhall Motors went on to be a major supplier of military trucks during World War II.*

MACK

1930s-1940s USA

MACK E SERIES

Mack started building its E series trucks in the 1930s, but the series would continue with the introduction of several different models through to the early 1940s and well into World War II.

Above: *Mack E series trucks were built as conventionals as well as the semiforward control "COE" type seen here. All featured the stylish "vee-shaped" radiator grille with chrome trim.*

All conventional-style E series trucks had the same cab, with a one-piece windshield and high-crowned fenders. They also all had hydraulic brakes, used the "Archimoid" steering mechanism and were fitted with five- or 10-speed monoshift transmissions.

COE E models

The E series also included cab-over-engine styles—all of them with an E-style grille and the by now famous bulldog ornament. There were six COE E models, including the EEU and EHU, weighing in at 12,000lb (5,542kg), the EFU and EMU, weighing 14,000lb (6,349kg), and the EGU and EQU, weighing 16,000lb (7,256kg). Mack also introduced an EQ cab-over truck in the early 1940s with a gross vehicle weight of 23,000lb (10,431kg). These cab-over Es had bigger engines. (Most Mack trucks were available with proprietary Buda or Cummins diesel engines.)

The first trucks in the E series were heavy- to medium-duty four-wheelers, but they quickly gave birth to lighter-duty trucks, for example the ED model, powered by a 67hp (49.9kw) L-head six-cylinder Continental gasoline engine. The lighter-duty ED trucks were styled after the heavier-duty E models, but were less elaborate, for example, having no chrome.

Mack would continue to produce the ED series until 1944 and would make several for wartime military use.

Specifications

Country: USA
Year manufactured from: 1930
Engine: cab over, diesel, or gasoline powered
Transmission: five to ten-speed monoshift
Payload: variable
Applications: multiple
Special features: hydraulic brakes

ERF

🛠 **1930s UK**

ERF C-14

At the beginning of the 1930s, the future of truckmaking would lie with diesel. E.R. Foden came out of forced retirement (he had been pushed to retire by Foden board members) to prove it.

Specifications

Country: UK

Year manufactured from: 1930s

Engine: Gardner diesel powered

Transmission: four-speeds forward

Payload: six tons (6.1 tonnes)

Applications: off road, heavy duty

Special features: body directly mounted on frame

Right: *The ERF CI4 was an assembled truck, built from components bought in from Gardner, David Brown, Kirkstall, Rubery Owen, Lockheed, and Clayton Dewandre. This wartime example has the Jennings "Streamline" cab introduced in 1938.*

Late in 1933, Foden and his newly formed company E.R. Foden and Son Diesel unveiled their first diesel truck—a two-axled six-tonner they called the C-14.

The truck had a Gardner 4LW automotive engine, a David Brown gearbox, axles from Kirkstall Forge, a chassis frame from Rubery Owen, Lockheed hydraulic brakes, and a Clayton Dewandre vacuum servo. It had a straight-grained ash frame for strength, aluminum panels which were light but also resisted corrosion. It had steel mudguards, a varnished interior, and wire-framed seats. It also had a stick and slat roof covered in leather cloth and plywood floorboards and the body was directly mounted on the frame.

Assembly truck

Assembly from purchased components was a departure from what other British truck companies were doing, because Leyland and Foden were still making their own parts.

But for E.R. Foden and Son Diesel, it worked. As their diesel trucks gained a reputation for strength and versatility, more models would follow. By 1937, the C14 (now known also as the No. 63) had been joined by a 6x4, a 8x4, and a twin-steer 6x2, and the company was soon providing trucks for big fleet buyers. The company grew so quickly that by 1935 it had to change its name to ERF to avoid confusion with Foden, the company which E.R. Foden had founded.

DIAMOND T
1930s USA

DIAMOND T200

Founded by C.A. Tilt, the Diamond T Company started out building passenger cars. In 1911, however, one of its customers wanted a truck and finally Tilt focused on utility vehicles.

Left: *With the Model 200, Diamond T produced a modern truck with pneumatic tires, spoke wheels, and a complete electrical system. The old wooden cabin was still used, but in 1933 a revolutionary new steel cab with V-type windshield was proposed.*

During the next two decades, the Diamond T ("T" for "Tilt") was an undisputed giant of truck manufacturing. From 1925 to 1930, the Diamond T company was able to offer a great range of models, from the "small" 1-ton (1.02-tonne) model to the heavy-duty 12-ton (12.19-tonne) truck. Most of these trucks were powered by Hercules or Hinckley engines and in 1926 new features such as steel spoke wheels and electric lights became standard.

Developments

In 1928, with the introduction of a new six-cylinder engine, hydraulic brakes, and an updated design using chrome, the Diamond T range was a real commercial success. In 1930, the 1-ton (1.02-tonne) Model 200 took the place of the Model 151. The new commercial truck was offered in two variants: one (Model 200) with a 128in (325cm) wheelbase with a new Buda four-cylinder engine rated at 50hp (37kw), and another (Model 215) with a 135in (342cm) or 158in (401cm) wheelbase powered by a bigger Hercules six-cylinder engine rated at 61hp (45kw). The four-cylinder Buda engine was very attractive and modern with a 3in (7.6cm), five-bearing crankshaft, which offered smoothness and reliability. The side-valves were totally enclosed and entirely water jacketed.

Specifications

Country: USA
Year manufactured from: 1929
Engine: four cylinder or six cylinder, 50 to 61hp (37kw to 45kw)
Transmission: four-speed
Payload: one ton (1.02 tonne)
Applications: stake body, delivery truck
Special features: new front styling with chromed radiator shell

BERLIET
1930 France

BERLIET DIESEL-POWERED TRUCK

Marius Berliet started experimenting with automobiles in 1894. His first car was built in a shed. Because he had to be able to get it out through the garden gate, it could only be as wide as that.

Above: *Berliet introduced its first diesel engines in 1930. They were of the indirect injection type.*

Specifications

Country: France
Year manufactured from: 1930
Engine: diesel powered Martin
Transmission: chain drive
Payload: not available
Applications: ideal for long haul
Special features: France's first diesel-powered truck

Berliet produced his first truck, for a silk manufacturer, in 1897. The first trucks were built on car chassis, but they mounted some of the biggest engines of the time—as much as 80hp (59.6kw). They were also forward control. Berliet eventually received the ultimate recognition as a truckmaker, producing 40 CBA trucks per day for the French military during the battle of Verdun in 1916. In 1917, both Berliet and Renault produced a tank specifically ordered by the French government, known as "the Diamond brand." During the war, Berliet produced up to 16 of these tanks every day at its plant in Venisseux.

Introduction of diesel-powered trucks

Berliet introduced France's first diesel-powered truck in 1930. It had a Martin diesel engine (Martin was a French manufacturer that had produced its first diesel engine in 1928). The new engine certainly stimulated sales because diesel was much less expensive and easier to maintain than gasoline. It also proved more powerful—ideal for the long-distance transportation of goods.

GMC

1930s USA

GMC HEAVY-DUTY TRUCKS

By the early 1930s, GMC trucks were available with payload capacities of up to 15 tons (13.6 tonnes). Unlike their colleagues at Chevrolet, GMC engineers were focusing on designing heavy-duty trucks.

Specifications

Country: USA

Year manufactured from: 1930s

Engine: six cylinder, gasoline powered

Transmission: five speed, over- and underdrive

Payload: up to 15 tons (15.2 tonnes)

Applications: multiple

Special features: first sleeper-cab optional

Tandem rear axles were already available on trucks like the 10-tonner T0-956 prior to 1933. And by 1933, GMC had fitted its first sleeper cab on a heavy-duty truck, showing the company had a pretty good sense of where trucking was going in terms of long haul. The same year saw downdraft carburettors as well as overhead valve engines adopted in the company's 2-tonner series.

By that time, GMC's heavy-duty trucks were coming standard with 707ci (11.6-liter) six-cylinder engines rated at 173hp (12.9kw). The engine used twin-plate clutches and had a five-speed transmission, with both over- and underdrive. Chain drive or worm drive was the standard on the tandem axles. And for the heaviest trucks, air brakes made by Westinghouse were standard as well. A six-cylinder, 70hp (52.2kw) side-valve engine, or the larger 120hp (89.5kw) overhead-valve six-cylinder engine were available for trucks with a payload capacity greater than 1.5 tons (1.52 tonnes) in 1934.

Streamlined design

The company redesigned their cabs in 1935, making even the heaviest-duty truck cabs more streamlined. Hydraulic brakes were also added, with full air actuation for the heaviest trucks and hydrovac for medium duty. By 1937, the company had 12 conventional models and 11 cab-over-engine models in its heavy-duty truck line. Diesel engines were standard for the 2- to 6-ton (2.03-to 6.1- tonne) range by 1939, unlike similar models built by Chevrolet.

Above: *GMC heavy trucks of the early 1930s were of very staid appearance but starting in 1936 the company adopted a more streamlined approach with stylish radiator grilles and sloping "vee'd" windshields.*

HANSA-LLOYD

1930s Germany

HANSA-LLOYD MERKUR TYP IIID

Hansa-Lloyd Merkur trucks were used mainly by the German army for the construction of telephone/telegraph lines, but the chassis were also commercially available.

Above: *Hansa-Lloyd trucks were the forerunners of Borgward. The Merkur was a four-tonner.*

Specifications

Country: Germany
Year manufactured from: 1930s
Engine: six cylinder diesel powered
Transmission: four-speeds forward, one back
Payload: variable
Applications: military construction
Special features: fuel consumption 10 miles per gallon (28km per liter)

The Typ IIID Hansa-Lloyd Merkur had a Klöckner-Humboldt-Deutz six-cylinder diesel engine that produced 75hp (55.9kw). The chassis was 22ft 6in (6.8m) long, 7ft 5in (2.25m) wide, and 8ft 6in (2.6m) high. Its wheelbase was 14ft 9in (4.5m). The truck had four-speed forward and one-speed back transmission and hydraulic brakes, as well as leaf-spring suspension. It weighed 9,040lbs (4,100kg) fully loaded. The fuel tank could hold 0.2 gallons (1 liter) and fuel consumption was approximately 28 miles per gallon (10km per liter).

By the time Hansa-Lloyd Merkur trucks were in production, the company for which they were named was already operating under another name. The company had been renamed Bremer Kuhlerfabrik Borgward & Company in 1921 by Carl Borgward, the company's partner.

First commercial vehicle

In 1924, the company brought out a three-wheel transporter powered by a 72ci (120cc) DKW rear belt-driven engine. It was first used in Borgward's own factory and then sold mostly to small tradesmen. In 1925, Borgward brought out the Goliath Rapid—a three-wheeled commercial vehicle which proved to be very successful.

LEMOON

🛠 **1938 USA**

LEMOON 6X4 TRACTOR

The Lemoon truck company started vehicle production as Nelson-Lemoon in Chicago in 1910. Company cofounder A.R. LeMoon went on to win a commercial vehicle trial with the company's first truck in 1911.

The winning vehicle weighed 1 ton (1.016 tonnes), and had a four-cylinder gasoline engine with double chain drive. Encouraged, Nelson-Lemoon brought out a 1½-tonner (1.52 tonner) in 1912. The next year, 2- and 3-ton (2.03- and 3.04-tonne) models were produced based on the same design. Not much would change about the vehicles for quite some time—the trucks were powered by Continental engines and this wouldn't change until the 1930s.

In 1927, the company's name was changed to Lemoon, and by the early 1930s, the company was offering straight-eight engines by Lycoming or Cummins diesel engines. By 1936, Lemoon models ranged from 2 to 12 tons (2.03 to 12.2 tonnes), and the cab-over-engine design had been introduced. (An articulated truck-tractor was also introduced that year.) In 1937, Lemoon made the switch to Waukesha six-cylinder 73 to 127hp (54.4 to 94.7kw) engines, which would power all eight of the models produced that year. That changed in 1938, when Lemoon offered engines built by Caterpillar, Cummins, Waukesha, and Continental. By then the cab for the Lemoon truck tractor had become more streamlined.

Above: *Le Moon's 6x4 tractor of 1938 was their heaviest model seen here hauling an excavator on a low-loader trailer.*

Specifications

Country: USA
Year manufactured from: 1938
Engine: six cylinder, gas powered
Transmission: dual chain drive
Payload: up to 12 tons (12.2 tonnes)
Applications: multiple
Special features: cab-over-engine design

DAIMLER-BENZ

1930s Germany

MERCEDES-BENZ GASOLINE-POWERED L SERIES

In 1924, Benz & Cie and Daimler-Motoren-Gesellschaft were selling Mercedes and Benz brands jointly. The association was followed by a merger into a new company called Daimler-Benz AG in 1926.

Above: *Mercedes Benz truck designations were prefixed "L" for "Lastwagen."*

Specifications

Country: Germany
Year manufactured from: 1930s
Engine: gas or diesel powered, four or six cylinders precombustion
Transmission: four-speed
Payload: up to 5 tons (5.08 tonnes)
Applications: multiple
Special features: four-wheel drive available

The two companies started streamlining product ranges and restructuring their plants. Benz would dominate the production of commercial vehicles within the new company. The Daimler plant in Berlin-Marienfelde would handle repair and parts, and only produce some custom off-road vehicles with four-wheel drive for a brief period during the 1930s.

First trucks

The newly merged company unveiled its first trucks at an international car show in Cologne, in France, in 1927. The Mercedes-Benz L1 through to L 5, N 1 and N 2 ranged in payload capacity from 1.4 to 4.9 tons (1.5 to 5 tonnes).

The L1 was the lightest, at 3.4 tons (3.5 tonnes). It had a four-cylinder 3.7-liter (224.6ci) gasoline engine that could provide up to 45hp (35.5kw). The heavy-duty L 5 weighed about 9.8 tons (10 tonnes) and was powered by an 491.6ci (8.1-liter) four-cylinder gasoline engine, that provided up to 70hp (52.1kw). The L 5 had the world's first six-cylinder diesel engine for vehicles.

MERCEDES-BENZ
1930s Germany

MERCEDES-BENZ L SERIES

The Daimler-Benz diesel-powered OM 5 six-cylinder would, until the mid-1930s, only be installed in one truck—the heavy-duty N 56 three-axle truck, introduced with the first L series gasoline-driven trucks in 1927.

That would change, however, when Bosch started to mass produce injection pumps. Up to then, injection pumps had been produced by Mercedes-Benz itself, at a much slower rate. The pumps were essential to the functioning of diesel engines, to deliver an exact metered amount of fuel, under high pressure, at the right time to the injector. The diesel injector, unlike in a gasoline engine, injects the fuel directly into the cylinder or a prechamber connected to the cylinder. The advent of mass-produced diesel injectors, therefore, literally fueled diesel engine production at Daimler-Benz.

The LO 2000 of 1932 was Daimler-Benz' first lightweight production truck with a diesel engine. It had a gross weight of just under 4.9 tons (5 tonnes) and a payload capacity of 1.9 tons (2 tonnes). The diesel engine had made a significant breakthrough. Under its short hood, the LO 2000 featured the OM 59 four-cylinder engine with 230.6ci (3.8-liter) displacement and a 55hp (41kw) output.

New series includes semis

The new series at first ranged up to the L 5000 with a 5-ton (5.08-tonne) payload. The series also included semitrailer tractors. All of them were available with gasoline or diesel engines, but the diesel engines of this generation were clearly more desirable than the gasoline-powered ones. And Mercedes-Benz made no secret of this—the trucks with diesel-powered compression-ignition engines had the word "diesel" on the radiator.

Above: *Diesel power gained popularity in the 1930s.*

Above, top: *In the 1930s most Mercedes heavy duty trucks had long hoods with set-back front axles, providing good engine access.*

Specifications

Country: Germany
Year manufactured from: 1930s
Engine: diesel powered four and six cylinder
Transmission: chain or worm drive
Payload: up to 4.9 tons (5 tonnes)
Applications: multiple
Special features: indirect injection in prechamber

SCANIA

1930s Sweden

SCANIA-VABIS TYPES 1544 AND 1561

By the 1930s, Scania-Vabis had introduced engines adaptable to different types of fuel. The first engine to do so was the Hesselman, introduced in 1926, which cut operating costs for drivers, but had its own problems.

Above: *In the 1930s Scania offered indirect-injection diesels that could be adapted to run on other fuels including gasoline and benzyl.*

Specifications

Country: Sweden
Year manufactured from: 1930s
Engine: six cylinder diesel, gasoline or light bentyl alternative
Transmission: four-speed
Payload: varied
Applications: multiple
Special features: engines built to burn variety of fuels

For example, the Hesselman engine could only achieve adequate working temperatures under heavy loads. And it used up too many spark plugs, not to mention the fact that the exhaust fumes emitted by the engine during cold starts stank.

Scania-Vabis solved the problem by developing an in-house solution—its own engine line. The company's first six-cylinder, 120hp (89.4kw) precombustion diesel engine (with six cylinders), with a seven-bearing crankshaft, was introduced in 1936. The same engine was ordered in a gasoline version, or in a version burning light bentyl (75 percent gas and 25 percent ethyl alcohol). In both cases, the engine would produce 140hp (104.4kw).

Three years later, Scania-Vabis launched another engine family in four-, six-, and eight-cylinder models, available in both carburettor and diesel versions. Among numerous shared components were cylinder heads, pistons, connecting rods, bearings, and exhaust systems. This "modular" approach to engine-building gave Scania-Vabis customers many specification options and advantages. For example, engines could be given different compression ratios by changing pistons

The Great Depression

The shared components also made the trucks that much more attractive to owners when it came to ease of maintenance and repair. It was largely thanks to its forward-looking technology and concern for running costs that Scania-Vabis would survive its next hurdle, the Great Depression.

GENERAL MOTORS

1936 Germany

OPEL BLITZ

General Motors acquired the German motor company Opel AG in 1929. The Opel Blitz was the result of this venture, but with the rise of the Third Reich, the German government decided to nationalize the company.

Above: *Just as General Motors acquired Vauxhall in the UK so it did Opel in Germany. During World War II Opel Blitz military trucks were widely used by the German army.*

Specifications

Country: Germany
Year manufactured from: 1936
Engine: four cylinder gasoline 68hp (50kw)
Transmission: not available
Payload: 2.95 tons (three tonnes)
Applications: general-purpose military truck
Special features: 4x2 "S" Type, 4x4 "A" Type, Half Track "Maultier" Type

More than 100,000 Opel Blitz ("Lightning") were built during World War II. The figure is far from the 800,000 GMC 6x6, but the Blitz was at first planned as a civilian commercial truck. This light vehicle was born in 1931 with a 220-cu.in (3.6-liter) Buick engine (Buick was part of General Motors). The first series had a 1.72-ton (1.75-tonne) payload, which then increased to 2.46 tons (2.5 tonnes) and finally 2.95 tons (3 tonnes) with the most famous Blitz, known as the "S" type. This model made its initial test in 1936 and looked modern with its steel cabin. The gasoline engine was far superior to any of its diesel competitors. After several testings, the Blitz was chosen by the German army. It could accept more than the mentioned payload and its ability to go where no other truck with two-wheel could go is legendary.

Variants

In 1940, Opel built a second version, a 4x4 truck known as the "A" type. About 25,000 of this cross-country Blitz were made and another half-track model was derived from the original frame, codenamed "Maultier" ("Mule"). Only 4,000 entered production, starting in 1942, and most were sent to the Russian front. The Blitz could be customized with various bodies, such as ambulances, tankers, radio-trucks, and general-purpose.

VOLVO

1931 **Sweden**

VOLVO LV66

With the LV 66, the Swedish Volvo company definitively entered the truck market. This truck was indeed a completely new commercial vehicle, designed to compete with the other local truck builder, Scania-Vabis.

Above: *The LV66 took Volvo into a heavier weight class in 1931.*

Specifications

Country: Sweden
Year manufactured from: 1931
Engine: six cylinder gasoline powered 56kw (75hp)
Transmission: not available
Payload: 3.4 tons (3.5 tonnes)
Applications: delivery truck, tanker, dump truck
Special features: first real Volvo truck, new engine DC 75, new gearbox, and new axle configuration

Until the beginning of the thirties, Volvo used some car parts in the construction of their trucks. As they began considering increasing the payload of their vehicles, however, the Volvo managers Assar Gabrielsson and Gustaf Larson decided to launch a larger truck with heavy components.

Power problems

From its first appearance on a drawing board, the new truck had to choose between a six- inline or an eight-inline cylinder engine with overhead or flathead valves. Finally, a six inline overhead valves engine was chosen. The gasoline-powered DC 75 engine offered 75hp (56kw) and ran in combination with a new four-speed gearbox. The new LV 66 and 68 ranges were sold from 1931. Different wheelbases and two- or three-axle configurations (from 1933) were available. For the first time, Volvo was also able to sell with a single or a double reduction. In 1933, these ranges could be powered by a new engine, the "Hesselman" engine, based on the gasoline engine, but which could run on diesel fuel thanks to a mechanical injection system. The LV 66 and 68 series were quite popular, and produced reasonable sales figures, but as three-axle trucks or as heavy-duty vehicles, the Volvo DC 75 engine wasn't powerful enough.

GUY

1931 UK

GUY 40/50 CWT

The Guy chassis was produced by Guy Motors of England. Founded by Sydney S. Guy in 1914, the company had immediately begun making 30 cwt lorries and in the same year produced an open "mail car."

Above: *This six-wheeler was the biggest 1931 Guy able to carry 11 tons (11.2 tonnes).*

In 1919, Guy introduced its "charabanc" (from the French for "wagon with benches"), along with the Guy Saloon (sedan) car. But the 1920s was a bad time for car production, so Guy decided to focus on trucks and buses. The company developed Britain's first dropped-frame chassis in 1924, ideal for ease of loading in both truck and bus functions. Two years later the company sold 170 of these chassis for bus and truck use to the Rio de Janeiro Tramway and Power Department.

Several trolley and bus designs would follow, but Guy continued to develop chassis that could be used for both trucks and buses. They included optional pneumatic tires and gasoline or Gardner diesel engines, low-loading chassis and forward control.

Chassis used for trucks and buses

But the Guy 40/50 cwt truck chassis produced in 1931 was used for both commercial and passenger purposes. The commercial range included ¾-, 5-, 6-, and 11-ton (0.76-, 5.08-, 6.1, and 11.2 tonne) versions. For commercial applications it included a 12-month guarantee coupled with two years of free periodical inspections.

The Guy chassis was marketed as the lowest-priced all British 40/50 cwt on the market. The company was clearly in competition with American manufacturers, who were just beginning to offer inexpensive, modern trucks to the British truck buyer. They made a wide range of light and heavy trucks for another three decades until it was taken over by Leyland in 1966.

Specifications

Country: UK
Year manufactured from: 1931
Engine: gasoline or diesel optional
Transmission: four- and five-speed
Payload: 11 tons (up to 11.2 tonnes)
Applications: multiple
Special features: lowest priced British-made truck on market

SISU
1931 Finland

SISU THREE-TONNER

Finland is a small country and therefore from the earliest days it imported most of its motor vehicles. As in many other small European nations, it was less expensive to buy an import than pay for a home-grown vehicle.

Specifications

Country: Finland
Year manufactured from: 1931
Engine: four cylinder petrol
Transmission: four-speed
Payload: three tons (3.04 tonnes)
Applications: variable
Special features: well suited to rugged and cold Finnish environment

Above: *Sisu trucks were uniquely the rough-forested regions, icy roads, and extreme cold of their Finnish winter environment.*

Up until World War II, most trucks in Finland were made in the U.S. But there were exceptions, and Sisu trucks were among them. Sisu would become Finland's first truckmaker, producing its first three-tonner in 1931. Sisu trucks were uniquely suited to their environment—the rough-forested regions, ice roads, and extreme cold of the Finnish winter. Thus the company survived long enough to be commissioned to produce military vehicles for the Finnish army during the Soviet invasion in 1939.

Merge with Vanaja

Sisu went on to merge with another Finnish truck manufacturer, Vanaja, in 1943. By then Sisu was focused on producing military vehicles and trucks, a priority in a small country caught between the war in Europe and the ever-present Soviet threat. The company was renamed Yhteissisu.

Yhteissisu experienced ongoing adversity —transportation equipment is naturally very vulnerable in wartime and most of the trucks that managed to survive the war were used to the point of destruction and then scrapped during the postwar years. Strict import restrictions and the regulation of foreign currencies favored vehicles made in the Soviet Union and East-bloc countries. But despite these difficulties, Sisu would grow, and in 1948 Yhteissisu was demerged and Sisu and Vaneja became independent companies once more.

FAGEOL

1929 USA

FAGEOL 10-TONNER

By 1929, Fageol trucks would come standard with complete electrics as well as full-pressure self-lubrication and pneumatic tires. The year 1929 saw the introduction of four-wheel hydraulic brakes.

Above: *This is one of only 10 all-aluminum trucks built by Fageol in 1929.*

The model range for 1929 included a 10-ton (10.1- tonne) six-wheel truck with an optional six-cylinder Hall-Scott engine, ideal for long hauls over the extremely rugged terrain and steep gradients of the Northwest Pacific states where Fageol trucks sold best. Other models, ranging from 2- to 6-ton (2.03- to 6.1-tonne) payload capacity were powered by Waukesha four-cylinder engines, with transmission and engines manufactured by Fageol.

Comfort was not the Fageol truck's strong point. The cab of a Model 135 in 1930, for example, was not exactly geared toward driver coziness. There was shelter in the form of a steel cab and roof, glass windshields, and windows were available, and left-hand steering was standard, but the only luxury was a leather-covered seat.

The trucks were still very expensive, but worth it, at least according to those who needed trucks that could do the job. Fageol was often the truck of choice for log haulers in the region. That changed with the onset of the Great Depression. Fageol tried to merge with Moreland of California, but the deal fell through, in part because of Fageol's extensive losses. By 1932, the company had gone bankrupt. It was reorganized as the Fageol Truck and Coach Company in 1932.

Specifications

Country: USA
Year manufactured from: 1929
Engine: four or six cylinder gas powered
Transmission: four-speed and worm drive
Payload: 10 tons (10.1 tonnes)
Applications: heavy duty long haul
Special features: four-wheel hydraulic brakes

KENWORTH
1935 USA

KENWORTH'S FIRST DIESEL TRUCK WITH SLEEPER CAB

In 1929 Harry Kent succeeded E.K. Worthington as president of Kenworth. Under Kent the company continued to grow, until the Great Depression hit. Kenworth responded, however, by creating new products.

Specifications

Country: USA
Year manufactured from: 1935
Engine: diesel powered four cylinder
Transmission: four-speed and worm drive
Payload: variable
Applications: long haul
Special features: first American factory to make diesel standard

Above: *While gasoline engines enjoyed greater popularity in North America where fuel was less expensive, Kenworth began offering a Cummins diesel alternative as early as 1933.*

In 1933, Kenworth would be the first American truck manufacturer to install a diesel engine as standard equipment in a truck. The first truck to receive a diesel engine would be a former gasoline-driven vehicle retrofitted with a 100hp (74.5kw) Cummins HA-4 diesel engine. From then on all of Kenworth's gasoline trucks could be retrofitted with diesel.

With the diesel retrofit came a change of venting systems as well—exhaust fumes would be streamed up over the cab. It was Kenworth engineer John Holstrom who came up with the idea of redirecting truck pipes this way. He did it because he noticed the fumes were coal-black if the diesel engine's fuel injectors were even slightly out of adjustment, and therefore very hard to see through when the driver was changing lanes. Customers loved the diesel retrofit, especially because diesel was so much less expensive.

First sleeper cab

In 1933 the company also brought out its first sleeper cab. Truck routes were getting longer and Kenworth recognized the advantages of providing truckers with somewhere to sleep, if only for a few hours. Yet again, Kenworth had successfully targeted a newly developing market.

STUDEBAKER
🛠 **1933 USA**

STUDEBAKER S SERIES

Studebaker started in 1852 as a family-run business building horse-drawn-wagons. It wouldn't be long before the company started building trucks. The company's first trucks were electric and designed by Thomas Edison.

Above: *The Studebaker S-series featured carlike styling typical of 1930s America. The company limited its truck range to light- and medium-weight models like the 3-tonner shown here*

Studebaker's first electric commercials, which could haul payloads of a couple of hundred pounds, sold well (for that time); in 1902, the year they were introduced, 29 of them were purchased. Later on in that same year, the company would bring out larger models with payload capacities ranging from ½ to 4 tons (0.5 to 4 tonnes). The company would continue to produce electric trucks and chassis until 1913.

Introduction of gasoline engines

In 1913, Studebaker unveiled a ¾-ton (0.76-tonne) payload straight truck with a gasoline-powered passenger-car engine. Not surprisingly, the vehicle had full electrics.

In 1928, the company acquired Pierce Arrow, and started building 1½- and two-tonners. But truck production stopped in 1929, when the market crash forced Studebaker to sell its Pierce Arrow stock. The company subsequently tried to buy the White Motor Company but failed. After a series of similarly unfortunate incidents, the company went into receivership in 1933, and the company's president, Albert Erskine, committed suicide.

New truck line unveiled

Still, thanks to the Pierce Arrow stock sale, the company recovered some of its losses. Its new managers Harold Vance and Paul Hoffman had one million dollars (made from the stock sale) to rejuvenate it. Vance and Hoffman started in 1933 by spending a whopping $100,000 on a print advertising campaign, just to let customers know Studebaker was still in the commercial vehicle business. That same year they brought out the new S model series of trucks.

Specifications

Country: USA
Year manufactured from: 1933
Engine: six cylinder gasoline powered
Transmission: four-speed
Payload: up to four tons (four tonnes)
Applications: multiple
Special features: decade-old proprietary engine

ZIS-5

1933 USSR

ZIS-5 4X2 THREE-TON CARGO TRUCK

About a million ZIS-5 cargo trucks were produced from 1933 to 1948. Together with the GAZ-AA trucks, the ZIS-5 trucks would become one of the two main Soviet trucks between 1930 and 1950.

Above: *Based on a contemporary American design the ZIS-5 had a six-cylinder gasoline engine.*

Specifications

Country: USSR
Year manufactured from: 1933
Engine: gas powered, six cylinder four stroke
Transmission: four-speeds twin reduction with bevel and spur gears
Payload: three tons (3.04 tonnes)
Applications: cargo hauling
Special features: shoe brakes on all four wheels

The ZIS-5 truck was gasoline-driven, had a carburettor, and was liquid cooled. The six-cylinder, four-stroke engine could produce up to 75hp (55.9kw). The truck was 19ft 1in (6.06m) long (bumper included) and 7ft 9in (2.35m) wide. It stood 7ft 1in (2.16m) high unloaded. Its rear axle clearance was 10in (25cm). The wheelbase was 12ft 6in (3.81m). The clutch had a twin plate, the gearbox had four speeds, and the differential type was twin reductor, bevel and spur gears. Steering was screw pin and suspension was leaf spring. The early ZIS-5s had cam brakes on all four wheels, which would be replaced by hydraulic brakes in 1949.

The ZIS-5 had a top speed of 37.2mph (60km/h). Unloaded, it weighed 6,835.5lb (3,100kg). The truck's fuel tank capacity was 13 gallons (60 liters). It could go nearly 124 miles (200km) on one tank, and required an oil change every 745 miles (1,200km).

The ZIS-5 was modeled after an American Autocar truck. It was built at the same plant as the AMO-15—the plant had been reorganized and expanded in 1929. In fact, when the AMO was rebuilt it was renamed ZIS after the plant's new name—the "Zavod imeni Stalina." By the summer of 1941 the Red Army would have 104,200 ZIS-5 trucks and by 1942, the trucks would be produced in three plants in Ulianovsk, the Urals, and Moscow.

Wartime trucks stripped down

Starting in late 1941 all ZIS-5 trucks would get flat fenders and some wooden parts (doors and seats, for example), and their bumpers would be permanently eliminated. Cost cutting during the war years would also see the trucks' right headlight eliminated. The truck would be nicknamed the "Tryohtonka" because of its 3-ton (3.04-tonne) payload.

DAIMLER-BENZ
1935 Germany

DAIMLER-BENZ L 6500 TO L 10000

The development of the Daimler-Benz L 6500, L 8500, and L 10000 diesel trucks could be linked to a 150 percent increase in commercial vehicle output in Germany previous to World War II.

Above: *The Mercedes Benz L10000 six-wheeler grossed 18.2 tons (18.5 tonnes) and was powered by a six-cylinder diesel engine mounted ahead of the front axle.*

Daimler-Benz not only increased its production of trucks, but also increased their technological development and range. Trucks that could carry larger payloads were added, as were bigger trucks, with more powerful engines.

Heaviest-duty trucks yet

The Daimler-Benz L 6500, L 8500, and L 10000 were the heaviest-duty trucks yet. They had three axles each. And with the hoods that appeared to go on forever, much longer than any to date, they would increase the Daimler-Benz truck range. (The giant L 10000 solo truck could weigh as much as 18.2 tons [18.5 tonnes] fully loaded.) They were powered by large six-cylinder inline diesel rather than gasoline engines, and with their displacement of 758.7ci (12.5 liters) they could provide up to 150hp (111.8kw).

Their introduction was none too soon—transportation distances had, by the mid-1930s, lengthened a great deal. But development wasn't confined to making bigger and more powerful engines. Design was advancing too. In 1938, a rounded short-nose cab design with a one-piece windshield replaced previous more angular long-nosed designs. (The new look lasted for quite some time, even after Daimler-Benz recovered from World War II and well into the 1960s.) Production methods changed as well, thanks to the huge demand.

Starting in 1936, a new generation of light trucks—the L 1100 to L 2000—started to roll off the production line. The short-nosed trucks, which would be the forerunners of today's delivery vans (except for the noses), were powered with 260D diesel car engines. These too proved to be very popular, although they did not match the German government's new design specifications.

The military dictates production

In the late 1930s, the German government became increasingly dictatorial and they restricted the production of motor-vehicle manufacturers to just four basic truck models. This was part of the German Reich-appointed "Plenipotentiary for the Motor Vehicle Business" Colonel (and later General) von Schell's so-called "Schell Plan." Manufacturers were told what to produce, and Daimler-Benz was assigned trucks with payload capacities of 2.9-, 4.5-, and 5.9 tons (3-, 4 1/2-, and 6-tonnes). The Schell Plan also required the company to develop and produce vocational vehicles for off-road use as well as tractors. Essentially, the plan forced German vehicle manufacturers to meet the needs of the German armed forces first and foremost.

Commercial design suffers

Unfortunately for German industry, military vehicles would not and could not meet the needs of commercial operators. Gasoline engines were more suitable. So Daimler-Benz's light trucks, as well as the L 701, were powered by gasoline once again. And light truck production would predominate, as did the manufacture of vocational trucks for use both on and off the road.

The military forced Daimler-Benz to produce some unlikely as well as dislikable trucks. In the early 1940s, for example, the company was forced against its will by the Third Reich to abandon its own 3-tonner design and produce the Opel Blitz 3-tonner design instead.

Production of the Opel Blitz started at the Mannheim plant in 1944. The product was designated as the L 701. With its angular makeshift wooden cab, it did not reflect the elegant design esthetic Daimler-Benz had hitherto cultivated.

Postwar production crippled

Despite this, production of the Opel Blitz design did continue at Daimler-Benz plants for a couple of years after the war ended. Thanks to the Third Reich and the economic and industrial ravages of the war, most Daimler-Benz plants lay in ruins. Mannheim, Stuttgart, and Gaggenau had survived, but were occupied by the Americans and (the French occupied Gaggenau). The occupying forces did not allow German manufacturers to produce three-axle trucks, or engines of more than 150hp (111.8kw). To make matters even worse, parts, workers, and raw materials were in very short supply.

Specifications

Country: Germany
Year manufactured from: 1935
Engine: diesel powered, six cylinder in line
Transmission: not available
Payload: up to 9.85 tons (10 tonnes)
Applications: long haul
Special features: short-nose design

Below: *While the L6500 carried around 7 tons (7.1 tonnes) solo it could handle about twice that weight in drawbar form.*

ERF
⚒ **1935 UK**

ERF SIX-WHEELER

ERF development and sales had taken off since the company's inception in 1933. One of ERF's first six-wheeler chassis was the No. 315, delivered to Tess-Side Motor Transfer in 1935.

Above: *Originally badged "E.R.Foden," the trucks were rebranded "ERF" soon after introduction to avoid confusion with rival vehicles, Foden, built just down the road.*

Left: *After launching their first truck, a four-wheeler, in 1933, ERF expanded their range to include three- and four-axled rigids. This is a CI5.62 six-wheeler with Gardner 5LW diesel engine.*

After bringing out the C14 (otherwise known as No. 63—after E.R. Foden's then age) in 1933, with a Gardner 4LW engine, the company had brought out the C15 with a Gardner 5LW engine. In November 1934, the company delivered this truck with a three-way tipper body. And by 1935 ERF's six-wheelers, with two steered axles and one rear-driven axle, was unveiled. The first six-wheelers had a payload of 16 tons (16.2 tonnes) gross.

While Foden suffered from having lost its founder E.R., it would quickly be onward and upward for ERF. This was despite the 1934 departure of Ted Foden, E.R. Foden's nephew, from the company. But ERF would remain undaunted, and continue to add more models to its already burgeoning range, including the CI56I with Gardner 5LW engine, and the CI66I with Gardner 6LW engine. Also available was a six-wheel chassis with one steered axle and two driven axles and grossing 18 tons (18.2 tonnes). And then there was the C1562 with the Gardner 5LW engine.

First eight-wheeler

It was while building an 18-ton (18.2-tonne), 6x4 six-wheeler in 1935, that the customer who had ordered it changed his mind and asked for a 22-ton (22.3-tonne) eight-wheel model instead. It would be the first eight-wheeler ever built by ERF. More eight wheelers, the C1662 with a Gardner 6LW engine, and the C1681 with the same engine, would follow.

Specifications

Country: UK
Year manufactured from: 1935
Engine: diesel powered
Transmission: four-speeds forward
Payload: 16 tons (16.2 tonnes)
Applications: long haul
Special features: six-wheel chassis

VOLVO
1935 Sweden

VOLVO LV82

With a more modern design, the new Volvo LV 80 series was not only an improvement on the older LV 71 and LV 73 series, but was also a new trucking concept created to increase load capacity.

Above: *The popular Volvo LV 80 became very familiar on the Swedish roads.*

Specifications

Country: Sweden
Year manufactured from: 1935
Engine: six cylinder gasoline
Transmission: four-speed
Payload: 2.46 tons (2.5 tonnes)
Applications: stake body, delivery truck
Special features: forward engine location, aerodynamic body, more comfortable cabin

The LV 80 (and the bigger LV 90) was notable for its engine location. To increase payload, the six-cylinder powerplant was moved forward on the front axle in order to improve the back axle load distribution. The distance between the bumper and the back of the cabin (BBC) was reduced too so that the truck could be fitted out with a longer body. This design made the truck profitable, but the driver's environment was extremely comfortable too. The roomy and ergonomic steel cabin received a heater (which was an absolute luxury at this time) and the shape of the body was made more aerodynamic, with a rounded, more streamlined shape.

Multiple formats

The new LV series was available in a wide array of formats, including various engines, suspensions, wheelbases, axles, wheels, and tires dimensions. Each component could be combined to offer the best truck the customer needed. For instance, the LV 82 was a 13ft 6in (4.10m) wheelbase truck, in a 4x2 configuration, powered by the Volvo EC75 flathead engine with a total weight of 5.41 tons (5.5 tonnes). Capable of operating from light to heavy duty, the LV 80 series entered a new trucking era with a modular concept. In fact, the new range was so popular that it quickly took over from the not-so-old LV 71 and LV 73.

VOLVO

1935 Sweden

VOLVO LV 93

When the first LV 80 series was shown in 1935, there were in fact three new series: the LV 81, the LV 83, and the LV 90 series. The LV 90 was fitted with a bigger engine and had an increased carrying capacity.

With their rounded aerodynamic bodies and less protruding hoods, the LV 80 and LV 90 series proved popular for medium loads during the prewar era. The LV 90 was designed for heavy-duty applications. The gross weight was increased from 5.31 tons (5.4 tonnes) to 6.29 tons (6.4 tonnes), or 6.84 tons (6.95 tonnes) according to their wheelbases.

Engine types

The main difference with the LV 80 series concerned the engine. The LV 90 was fitted with an overhead valves engine instead of the flathead EC 75. The total horsepower was the same, but the OHV DC 75 was a bit more nervous. The customer could choose an economical Hesselman HA 75 engine too. In 1936 emerged a more powerful FC 90 engined (90hp/67kw) truck. It proved to be a reliable and tough vehicle, but like all the other trucks from its series it had a relatively short production life. The LV 80 range terminated in 1939, with the introduction of the "sharp-nose" truck series. Volvo always preferred to launch a new truck rather than feature an existing model with important modifications.

Above: *About 2,950 LV 90 were built between 1935 and 1939.*

Specifications

Country: Sweden

Year manufactured from: 1935

Engine: six cylinder gasoline 90hp (67kw)

Transmission: four-speed

Payload: 3.44 tons (3.5 tonnes)

Applications: stake body, delivery truck, tanker, platform

Special features: long hood, bigger six inline cylinder engine.

KENWORTH

mid 1930s USA

KENWORTH 516 "BUBBLENOSE"

Kenworth unveiled its first cab-over-engine design, the "Bubblenose," in 1936. The truck was designed with a shortened nose to provide maximum payload while complying with the overall state truck length limits.

Above: *Length limits imposed in some states during 1935 led to an increase in cab-over-engine trucks. Typical of the era was the Kenworth 516 "Bubblenose" 6-wheel truck seen here with a drawbar trailer.*

Specifications

Country: USA
Year manufactured from: 1930s
Engine: four or six cylinder gas powered, four cylinder diesel powered
Transmission: five-speed
Payload: up to four tons (3.6 tonnes)
Applications: multiple
Special features: chrome grille

Weight restrictions had also been imposed by the Motor Carrier Act in 1935, so Kenworth had also begun building its own cabs from sheet metal. Chrome was also an esthetic element in the overall Kenworth design; the trucks were instantly recognizable thanks to their chrome grille.

Kenworth partner Harry Kent would not live to enjoy the new design's success. In 1937, he died of a heart attack. Nevertheless, Kent's legacy of cost-cutting and manufacturing efficiencies would continue, resulting in Kenworth becoming increasingly successful over the next three years.

Ten truck models

Among them would be 10 models of trucks, including the 2-ton (2.03 tonne) Model 88, powered by an 83hp (61.9kw) Hercules JXC engine and the 7-tonner Model 241C powered by a 125hp (93.2k) six-cylinder Cummins diesel engine. Buda and Herschell-Spillman gasoline engines were also available for the 4-tonner model 146B, and the 7-tonner Model 241A. (All gasoline engines had six cylinders by 1937.) The 505, 506, and 507 models were powered by Cummins four-cylinder diesel engines.

Kenworth also got into bus production at the end of the 1930s. The company custom-built the canvas-topped Tri-Coach bus, which could carry 18 passengers, for the Rainier National Park Company. Five of them were delivered to Rainier in 1938. Small buses were also produced. But truck production would continue to be Kenworth's focus, with at least 17 six-wheeled truck models available by 1940.

ERF
1937 UK

ERF OE4

The ERF OE4 weighed only 3 tons (2.72 tonnes) unloaded, but was capable of hauling a 6-ton (5.4-tonne) payload. Like most commercial vehicles of its size built in the UK then, it had a frame swept up over the rear axle.

Specifications

Country: UK
Year manufactured from: 1937
Engine: four cylinder diesel powered Gardner
Transmission: five-speed
Payload: six tons (6.1 tonnes)
Applications: multiple
Special features: could haul twice its weight unloaded

Above: *The OE4 was ERF's lightest truck and was powered by a Gardner 4LK diesel which boasted excellent fuel economy.*

The OE4 had other distinctive design elements as well, thanks to the ingenuity of its designer Ernest Sherratt. The stylish grille was positioned in front of the radiator so as to conceal it. This look was very modern for its time and it would be used for many future ERF trucks in years to come.

The Gardner diesel engine would be used to power the small but powerful OE4. In the late 1930s, Gardner would introduce its 230.6ci (3.8-liter) four-cylinder 4LK diesel engine, capable of up to 57hp (42.5kw) at 2100rpm. The light power unit, with matching transmission, was ideal for the ERF OE4.

By the beginning of 1939, an estimated 40 ERF trucks were rolling off of the production line every month. But six months later, World War II would have its effect on industrial production in the UK.

Design innovation continues

The company did, nevertheless introduce a few new vehicles at the start of the war, including its 7.5-ton (6.8-tonne) dump truck. Engineer Ernest Sherratt designed the truck with two goals in mind: weight reduction and low body-floor height. Weight was strictly controlled by careful selection and integration of parts.

FODEN

1937 UK

FODEN DG RANGE

It would take time to convince Foden's conservative board members that diesel power would overtake steamers. Several company officials were believers, among them E.R. Foden, his son Dennis, and brother William.

Above: *In 1937 Foden brought out its DG range, powered by Gardner LW series diesel engines with four, five, or six cylinders.*

Specifications

Country: UK

Year manufactured from: 1937

Engine: diesel powered, four to six cylinder Gardner

Transmission: five-speed

Payload: variable

Applications: delivery, then heavy duty

Special features: first Foden diesel trucks

Eventually, E.R. Foden was forced to resign. To Foden's eternal regret, E.R. founded ERF and in June 1933 brought out the diesel-powered C-14, which made the nascent company an almost instant success.

ERF's success and the fact that insurers were increasingly reluctant to insure steam boilers, finally convinced Foden board members that the internal-combustion engine was in fact necessary to the survival of company's commercial vehicle line. Thanks to some of the company's more forward thinkers, Foden had already begun experimenting with gasoline-driven vehicles. Foden then used several gasoline engines before focusing on diesel trucks such as the R and S-types.

DG range

Finally the company convinced William Foden to come back from Australia. His return heralded the production of the company's first diesel trucks and turned their fortunes around. In 1937 Foden brought out its DG range, powered by Gardner LW series diesel engines with four, five, or six cylinders and with a bore capacity of 84.9ci (1.4 liters). Foden first used these engines in trucks, but their use in heavy-duty commercial vehicles such as ballasted drawbar tractors followed.

FWD

🛠 **1937 USA**

FWD COEs

According to some truck historians, FWD was the first North American truck manufacturer to introduce the cab-over-engine design to its commercial vehicle range.

Left: *In 1937, FWD brought out a COE cab with a two-piece V-shaped windshield. It would be followed by the four-door COE crew cab. Soon COEs would be available in a wide variety of shapes and sizes, even on 6x4s.*

The company had its start in 1910, when Otto Zachow, a machinist in Clintonville, Wisconsin, patented a double-Y universal joint in a ball and socket. Zachow's U-joint allowed for the wheels of a vehicle to rotate and be steered simultaneously. With the help of his brother-in-law William Besserdich, who was also a machinist, Otto subsequently built a steam-powered passenger car, which used the new U-joints on its axles.

The steamer wasn't much of a success, so the brothers-in-law turned their hands to gasoline-driven vehicles. Their very first, nicknamed the Battleship, weighed 3,800lb (1,723.3kg). But the brothers still had no intention of getting into automotive manufacture. They just used the vehicle to demonstrate their U-joints. Finally, a group of investors managed to encourage the brothers-in-law to start their own company building four-wheel-drive vehicles. The Four Wheel Drive Automotive Company (FWD) was founded in 1909.

FWD dominates military production

World War I proved profitable for FWD, with the American military ordering so many FWD trucks that other manufacturers were given special licenses to build them as well. Postwar, with half the market flooded with FWD trucks, the company would get by selling spare parts for their maintenance.

Through the 1920s the company made several improvements to its truck line, including introducing pneumatic tires as standard equipment on trucks like the Model B, which had been produced since 1918. The company also focused on building special-purpose trucks, for fire services, utilities, oil hauling, etc.

By the mid-1930s, FWD had innovated further—expanding its product range in terms of payload capacity and powerplants. In 1937, FWD brought out a COE cab with a two-piece V-shaped windshield. It would be followed by the four-door COE crew cab. Soon COEs would be available in a wide variety of shapes and sizes.

Further innovations

FWD introduced new Cummins diesel powerplants as well as double-reduction or multiple-geared axles. With a short wheelbase and a shorter frame than a conventional truck, the COE series could maximize the payload capacity with a longer body or a longer trailer. But the drawback of such a cabin was the engine inside the driver's less roomy cab. The noise, the heat, and the burned-oil smell were uncomfortable, especially during the summer. FWD trucks retained the old-style front cabin during the thirties and didn't keep up with the "streamlined" fashions. The trucks were mainly used in severe-application duties, and the aerodynamic design was secondary. During World War II, FMW supplied the U.S. forces with military trucks. The SU heavy model was built as a COE or Conventional model for the U.S. Marine Corps, and for the Red Army and British forces for use as an artillery tractor. It was powered by a 115hp (86kw) Waukesha gasoline engine and could carry up to 5/6 tons (5.08/6.1 tonnes).

Postwar cabs

After the war, the company used a new cabin built by International, then a Dodge cab was fitted on the new Tractioneer series introduced in 1958. The company also developed custom-designed trucks to utilize Seagrave fire apparatus.

Specifications

Country: USA
Year manufactured from: 1937
Engine: variable
Transmission: five-speed and worm drive
Payload: variable
Applications: multiple
Special features: two piece V-shaped windshield

Left: *FWD Sucoe all-wheel-drive trucks came into their own as military vehicles in World War II.*

VOLVO

⚒ **1937 Sweden**

VOLVO TVA

As World War II grew closer, European countries were forced into defensive thinking. Sweden was no exception, and Volvo engineers were ready and eager to produce heavy-duty military vehicles.

In 1937, Volvo would unveil a design for a heavy-duty military vehicle customized for off-road use. The new TVA was a radical departure from any Volvo truck hitherto produced. It was a rugged 6x4, intended originally for towing artillery pieces and antiaircraft guns. It had an extra two smaller wheels (besides the six) between the first and second axles, which reduced pressure on the front two axles when the vehicle threatened to sink into softer ground. The TVA was powerful, with a two-axle drive, and strong tires made of thick rubber with off-road treads.

It had a large gasoline engine, similar to the one already used in the LV29 Sharpnose trucks, but with a sump pump added to ensure lubrication even during steep climbs. The frame of the TVA was tubular, with axles independently attached. This made the vehicle hold the ground better, even though the front axle wasn't a drive axle.

Little interest outside Sweden

Volvo would try to sell the TVA to other European countries, but there was little interest, probably because these countries already had their own truck industries. The vehicle proved popular with the Swedish armed forces, however, so much so that production of the truck would continue through to 1940, when an only slightly improved version of the TVA—the TVB—would roll of the Volvo production line and replace it. As for the TVA, the truck is still in service in some areas of Sweden today.

Above: *The TVA was intended for towing artillery and antiaircraft guns. It had an extra two smaller wheels (besides the six) which could be between the first and second axles.*

Specifications

Country: Sweden

Year manufactured from: 1937

Engine: gasoline powered with sump pump for lubrication

Transmission: five-speed

Payload: not available

Applications: towing artillery and antiaircraft guns

Special features: customized for off-road use.

FORD

1930s USA

FORD V-8 MODEL BB

In 1932, the new Ford V-8 caused a sensation because it was reliable, smooth, and powerful and had a low-cost flathead engine. It powered the new "BB" type commercial truck chassis that superseded the "AA" types.

Above: *There's no doubt about the engine with the "V-8" logo on the grille clearly displayed.*

Specifications

Country: USA
Year manufactured from: 1932
Engine: eight V cylinders, gasoline powered, 65hp (48kw)
Transmission: three gears
Payload: 1.52 tons (1.5 tonnes)
Applications: stake body, delivery truck, tanker, platform, dump truck
Special features: new V-8 engine

The "Flathead" Ford V-8 engine immediately became a success, despite initial casting troubles and overheating problems. With a cast-in-one piece, it had a 90 degree arrangement with a 221-ci (3.6-liter) displacement for 65hp (48kw). It wasn't really sophisticated, but Ford could offer eight cylinders for a price that no other manufacturer could. The new BB-type truck was offered in 131½in (334cm) and 157in (398cm) wheelbase with a new heftier steel frame. The semielliptic springs, an underseat fuel tank, and a revised shaft drive were fitted too. Despite the revolutionary engine, the Great Depression didn't help sales and Ford had to face up to million-dollar losses. But the fleet operators were very enthusiastic about the V-8 trucks, with their "V-8" logos proudly fixed to the front grille, and in 1934 the engine was improved with a new counterbalanced cast-alloy steel crankshaft, which contributed to increased smoothness by reducing vibrations. A new dual-downdraft Stromberg carburettor was fitted too at this time. In 1935, a new body shell was launched and the revised frame changed the name of the truck series. The "BB" type was phased out to become the Model 51.

VOLVO

1938 Sweden

THE VOLVO "SHARPNOSE" LV101

Volvo "Sharpnose" trucks were first introduced in 1938. Mostly light- to medium-duty vehicles, trucks in the Sharpnose series shared the same pointed appearance, which earned them their nickname.

Above: *The beaklike appearance of its radiator grille earned the Volvo LV101 the nickname of "Sharpnose."*

The very first of the series, the LV101, was the smallest truck ever built by the Swedish automotive company, and in terms of technology, very similar to the taxis the company had already built. (In fact, the chassis were almost identical.) Like later trucks in the first generation of "Sharpnoses," it had a 222.7ci (3.67-liter) side-valve EC engine rated at 86hp (64.1kw). The transmission was a three-to-four-speed mechanical unsynchronized gearbox and the chassis components were simple and rugged. LV101 and LV102 frames were arched over the rear axle to allow for a low platform, making loading and unloading easier. Later versions included a medium-duty truck suitable for heavier cargoes, which could also be outfitted as a small tipper. The largest Sharpnose was the LV110 (1949), with a choice of wheelbases, but unable to go long distances becauase of its 86hp (64.1kw) side-valve engine.

Foreign design influence

Volvo was not considered a major supplier of heavy-duty trucks at the time, lagging far behind the United States, England, Germany, and France in terms of modern design and sheer production volume. This probably explains why Volvo trucks of the mid- to late

1930s appeared to be heavily influenced by the design trends of the major truck producing countries. Not surprisingly, the engineers who designed them had, for the most part, trained and worked at American car and truck manufacturing companies. They had in common, therefore, a major preoccupation with body design, shared by management at the time. In fact, articles in magazines such as Volvo's customer magazine, *Ratten*, described the Sharpnose design as if it was a work of art, calling the chassis "attractive and elegant."

Producer-gas units

While early Sharpnoses were fairly simple, mechanically speaking—they all used a side valve engine of 75–86hp (55.9–64.1kw), when running on gasoline—the company also made trucks especially for wartime service that put out 50hp (37.2kw) running on Volvo's patented producer-gas units, which burned wood. Volvo developed and introduced the units in mid-September 1939, as national gasoline supplies rapidly depleted. The units, which were much easier to feed than units with coal-gas burners developed for other trucks and buses, only increased the popularity of Volvo trucks. The result was that Volvo's production capacity was stretched to the limit and the company could only produce a small number of trucks, buses, taxis, and cars at the same time.

The LV103 was introduced in 1943 at the behest of Swedish defense. It was the first Volvo truck to include a trailer. The company had previously specialized in constructing the chassis alone, allowing bodybuilders to complete the truck with a flatbed or trailer as required. But the single-axle LV103 tractor included a long semitrailer bed. The top speed of these trucks was limited, however, and made long hauls difficult because the engine developed high revs even at low speeds.

Some elements of the popular Sharpnose design would endure to make their presence felt in later designs over the years. In fact, the series survived for quite some time, with the LV101 only going out of production in 1945.

The LV110, a heavier version, remained in production until 1946 and the intermediate LV102 stayed in production well into the 1950s, with a wide range of applications. In fact, the LV102 was so popular with customers, who were willing to buy just about anything that was available on the meager postwar market, that Volvo brought out the LV105 series, which was basically a LV102 with a modernized engine (ED) and a new instrument panel.

Volvo started to become a major supplier of heavy-duty trucks on a global scale in the mid-1950s. Unfortunately for the Sharpnose design, it looked completely outdated.

Specifications

Country: Sweden
Year manufactured from: 1938
Engine: side valve gas powered
Transmission: four-speeds to bevel
Payload: variable
Applications: multiple
Special features: optional wood burning producer gas units

Below: *The late 1930s Volvo styling showed a strong American influence.*

GMC

1939 USA

GMC AFD COE

During the 1930s, the famous General Motors Company was more than a pioneer in the automobile industry. It was in fact a "mixed" corporation, producing cars, trucks, and buses.

Specifications

Country: USA
Year manufactured from: 1939
Engine: three or four cylinder diesel
Transmission: five-speeds to bevel
Payload: five tons (5.08 tonnes)
Applications: tanker, platform, stake body
Special features: two-cycle Detroit-Diesel engine

In 1925, GM bought a controlling interest in the Yellow Cab Company, a bus manufacturer that built all the GMC-labeled trucks until 1943, when GMC acquired Yellow Cab in its entirety. In fact, the GMC trucks were in competition with another GMC division, the Chevrolet Company, but this competition was looked on as a stimulus for innovation and profit. Finally, Chevrolet was ordered to build light trucks while GMC concentrated on heavier models. So in 1931, the GMC trucks were offered with up to 15-ton (15.2-tonne) capacity. The cab-over-engine design became popular, and in 1937 the company offered 11 COE models with gasoline engines rated from 85 to 110hp (63 to 82kw).

Diesel Engines

In 1939, GMC introduced its very first diesel engines. These were built by a new subdivision, Detroit-Diesel. These engines could only be used in trucks with up to 6 tons/6.09 tonnes. The new engines were two-cycle, three- or four-cylinder powerplants. The bigger conventional model was the T-61H rated at 10 tons (10.16 tonnes) and the more expensive COE variant was the F-61H. The COE, because of its smaller cab, could carry more payload (the F-61 had a 12-ton/12.2-tonne capacity chassis) and was very much appreciated by the truck operator, despite the fact that the engine accessibility was much less practical.

Above: *This 13,000lb (5,909kg) GMC Tanker was a 1939 model, one of the last prewar models. It kept its fine chromed grille, but as World War II began in Europe the company had to convert over to military production with a less attractive truck series.*

AUSTIN

1939 UK

AUSTIN K2

In 1939 the Birmingham-based Austin company entered the truck field with an extended range up to a 5-ton (5.08-tonne) payload capacity. The K2 was the most widely used truck.

Above: *Austin's new K range of light to medium trucks was first announced in 1939 but full production did not begin until after World War II.*

Right: *Such was the K model's similarity to its rival, the Luton-built Bedford, that it was nicknamed the "Birmingham Bedford."*

Specifications

Country: UK

Year manufactured from: 1939

Engine: six cylinder gasoline

Transmission: four-speeds to bevel

Payload: two tons (2.03 tonnes)

Applications: stake body, parcel delivery truck, dropside platform

The design of the Austin trucks was very similar to the Bedford trucks, and Austin was actually nicknamed the "Birmingham Bedford company." The 2-ton (2.03-tonne) K2 was designed before World War II, but the civilian steel cab model didn't get underway until 1945. During the war, however, about 13,000 K2/Y were built as military ambulances in Austin's Longbridge works.

Postwar service

After the war, the K2 began a new career as delivery van, cattle truck, or dropside wooden platform vehicle. The engine was a well proven gasoline six-cylinder rated at 27hp (20kw) with a four-speed gearbox. The Austin K2 and the bigger 5-ton (5.08-tonne) K4 were very popular on the British roads and they looked modern with a steel roof and a V-windshield. In 1947, a new commercial vehicle entered the range. It was the Three Way Van K8 light truck, which became famous with its forward-control cab and its steel body. At the beginning of 1950, all the conventional K series had a facelift, with more modern hoods and fenders. But they were phased out when Austin merged with the British Motor Corporation in 1951.

VULCAN
⚒ **1938 UK**

VULCAN TWO-TONNERS

The Vulcan Motor and Engineering Company saw its beginnings in Southport, Lancashire, in 1901. They would show their first gasoline-powered car at the Liverpool Cycle and Motor Show the following year.

Specifications

Country: UK
Year manufactured from: 1938
Engine: diesel or gas powered six cylinder
Transmission: four-speeds to bevel
Payload: two tons (2.03 tonnes)
Applications: commercial, municipal
Special features: low floor chassis for ease of loading and unloading

Above: *Vulcan 2-ton (2.03 tonne) forward-control trucks of the late 1930s were well engineered along the lines of heavy trucks.*

By 1906, the company was producing more than 100 vehicles per year. Vulcans were being sold outside Lancashire and even outside the British Isles—in New Zealand, Australia, India, Canada, Africa, Malaysia, Indonesia, Newfoundland, Ceylon, Spain, and the West Indies. Also in 1906, William Hampson, a cabinetmaker and joiner, became part of the company, later taking charge of the coachbuilding side of the business.

Passenger car production was suspended with the outbreak of World War I. From 1914 to 1918, Vulcan Motors made aircraft frames and engines. But the company still found the time to produce trucks as well (mostly for military use), leaving it in a much better position to continue with commercial vehicles production after the war.

Postwar, the company continued with both passenger and commercial vehicle production. Vulcan even developed a new six-cylinder gasoline engine in 1926, which it offered for sale on its own.

Commercial vehicles

In 1927, Vulcan switched its focus to commercial vehicles and entered the municipal vehicle market as well, by introducing a low-floor chassis. After the war times were tough economically worldwide, and during the 1930s the once-international company went into receivership.

SCAMMELL

🛠 **1939–1940 UK**

SCAMMELL PIONEER MILITARY SERIES

The first Scammell articulated truck was a 7.5-tonner built in 1921. From then on the company would focus mainly on building articulated and straight trucks with eight wheels.

Above: *The Scammell Pioneer's great strength was its exceptional off-road capability.*

Specifications

Country: UK

Year manufactured from: 1939

Engine: six cylinder Gardner diesel

Transmission: five-speed

Payload: not available

Applications: heavy duty, off-road, military

Special features: "breakdown" tractors fitted with folding cranes

In 1927, however, Scammell would come out with the Pioneer, a six-wheeled, off-road heavy-duty tractor with a rocking-type front axle and two feet of vertical movement at each rear wheel. The innovation made for incredible cross-country performance, which would make the Pioneer ideal for use in oil field and lumber operations.

With the outbreak of World War II, the Pioneer would also prove exceptionally useful for the British Army. The truck was used first as a tank transporter, but it would soon be put to other uses as well. The wartime Scammells were equipped with Gardner six-cylinder diesel engines, which gave them the high torque at low speeds they needed for tank and truck recovery operations. And they were also ideal for towing the heavy pneumatic-tired guns used at that time.

Winch or crane

To move guns across ground or to recover stuck vehicles, the pioneer artillery tractor was fitted with a chassis winch in 1939, which could lift 8 tons (8.1 tonnes). The first 43 "breakdown" tractors, designated SV1S or SV1T, were also fitted with folding cranes.

CHEVROLET
🛠 **1939 USA**

CHEVROLET ¾-TON MODELS JD AND JE

Before becoming part of General Motors in 1918, Chevrolet was an independent company founded in 1912 by William Crapo Durant. Chevrolet started building trucks in 1918.

The company's first trucks had payload capacities ranging from three-quarters to one ton and were unexceptional vehicles, with solid rubber tires and worm drive. They were powered by a 225ci (3.7-liter) four-cylinder splash-lubricated overhead valve engine made by the truckmaker itself. The trucks also used coil ignition, had complete electrics, and three-speed transmission.

They sold exceptionally well, Chevrolet selling 879 of them in the first year of production alone. By the end of 1919, another 7,300 were built. In the meantime, the company would bring out its first air-cooled four- and six-cylinder engines in the early 1920s. Water-cooled engines would follow, with the introduction of the 1-tonner Chevy trucks with bevel drive in 1923.

Improvements

By the 1930s, the truckmaker was using both six- and four-cylinder engines, and had made several improvements, including dual rear wheel options, stronger rear axles, and all-expanding brakes with the rear fully enclosed. Trucks were also fitted with an electric fuel gauge (they were still gasoline-driven at the time).

Above: *The JD model sported a two-piece V design windshield and a heavier grille.*

Specifications

Country: USA
Year manufactured from: 1939
Engine: four or six cylinder gas engines
Transmission: four-speeds to bevel
Payload: ¾ ton (0.76 tonne)
Applications: commercial delivery, long haul
Special features: hypoid rear axles

PETERBILT
1939 USA

PETERBILT 260GD

Peterbilt owed its beginnings to the California-based Fageol Truck Company, a truckmaker that specialized in building trucks for the market in the western United States.

Above: *Peterbilt adapted existing Fageol designs but achieved huge weight savings with extensive use of aluminum alloys.*

Specifications

Country: USA
Year manufactured from: 1939
Engine: diesel powered
Transmission: Cummins four-speed main and three auxiliary, double reduction at rear
Payload: not available
Applications: multiple
Special features: air brakes

Unfortunately, the Depression years would interfere with sales and Fageol would go into receivership in 1932. That was when T.A. Peterman, a lumberman from Washington, started worrying about where he could get good trucks. Peterman had been rebuilding surplus army trucks and modifying old logging trucks for use in his business. But by 1938, his lumber operations had expanded and he needed more trucks. He bought the old Fageol factory and started building custom chain-drive logging trucks in 1939. Of course, it would take some time for Peterman to truly reconcile the ins and outs of truckmaking.

Experiments with aluminum

For instance, the 1939 Model 260GD had an aluminum grille, which reduced weight but which also broke easily. So the use of aluminum in the grille would be abandoned after just one year.

The engine proved more successful, however. It was an HB6 Cummins diesel that could develop up to 150hp (111.8kw), with a four-speed Brown-Lipe main transmission, as well as a three-speed auxiliary one. The 260GD also came equipped with a Timken double-reduction rear axle and air brakes.

As with most truckmakers, Peterbilt's success or failure would depend largely on road development, industry, and military needs during World War II. While Henry Ford was cranking out hundreds of trucks a day, Peterman set his sights on building 100 trucks a year, concentrating on quality, not quantity. Peterman succeeded building 14 units in 1939 and 82 in 1940.

STUDEBAKER
1939 USA

STUDEBAKER M29 (UST24) "WEASEL"

Studebaker dropped its Cab-Forward series in 1941 to start production of its M series. The M3 was a light truck, one of several attempts by Studebaker to crack the pickup market.

Specifications

Country: USA
Year manufactured from: 1939
Engine: six cylinder diesel powered liquid cooled
Transmission: three-speeds forward one reverse rear axle chain drive
Payload: not available
Applications: military, specially created for Arctic reconnaissance and commando raids
Special features: the Weasel could float

Above: *The Studebaker Weasel tank was manufactured from 1942 to 1943, specifically for wartime use.*

With the introduction of the M series, Studebaker would also go back to using its own engines exclusively. (Previously Waukesha engines had also been available.) The Studebaker engines were six-cylinder diesels that could develop 94hp (70kw).

Introduction of the "Weasel"

The Studebaker "Weasel" tracked personnel and cargo carrier was manufactured from 1942 to 1943, specifically for wartime use. It was fully tracked at up to 20in (50cm) wide, had an engine positioned in front, two-speed rear axle drive, and was powered by a six-cylinder liquid-cooled Champion engine (an exception to the rule for the M series), which could produce up to 65hp (48.4kw) for a maximum speed of 38mph (58km/h). It was 10ft 6in (3.2m) long, 5ft 6in (1.67m) wide, and 5ft 11in (1.8m) high. It had a ground clearance of 11in (28cm). The vehicle was specially created for Arctic reconnaissance and commando raids, but was used much more widely by the American military, because it could also float. The M-29 evidently proved exceedingly useful for the military as an estimated 15,000 of them were produced by Studebaker in just two years.

VOLVO
1939 Sweden

VOLVO "ROUNDNOSE" RANGE

By the mid-1930s, Volvo had become the biggest truckmaker in northern Europe, and its product range had grown to include modern vehicles with highly efficient engines and huge payloads.

Above: *Like the earlier "Sharpnose," the "Roundnose" look was inspired by the designs of other truckmakers in North America, Britain, and Germany.*

Right: *In 1939 Volvo revised its front-end styling, giving the grille a more rounded appearance on this LV142.*

Specifications

Country: Sweden
Year manufactured from: 1939
Engine: gasoline, Hesselman or producer gas, side or overhead valve
Transmission: four-speeds to bevel
Payload: variable
Applications: mainly military
Special features: Roundnose design

The new "Roundnose" truck design was introduced in 1939, as a 1940 model. It would prove to be one of Volvo's most successful trucks of its time. The Roundnose series consisted of a wide range of trucks with a similar look but differing applications.

But there were subtle differences in design too. For instance, there were three different lengths for the front of "Roundnose" trucks. Two of them were available with gasoline, Hesselman, or producer-gas engines. The "Roundnose" look was inspired by the designs of other truckmakers in North America, Britain, and Germany.

The Roundnose design was introduced at the worst of times, at the very beginning of World War II. Civilian sales would therefore be few and far between. Fortunately for Volvo, military orders would make up the difference.

Popular postwar

The popularity of Volvo's Roundnose trucks would continue after the war, the most successful being the LV125 series. More than 10,000 of these would be sold. Roundnose trucks varied widely in terms of what they could do. They were for the most part powered by a basic side-valve engine providing up to 90hp (67.1kw). But gasoline engines with overhead valves providing up to 105hp (78.2kw) were also available. Also, while producer-gas engines were used during World War II, Hesselman engines, which burned cheaper oil fuel, were also another option.

WHITE
1939 USA

WHITE HORSE

By the end of the 1930s, the White Motor Truck Company was producing a wide range of commercial vehicles for the transportation of both passengers and goods, and was manufacturing its own engines.

Above: *White Horse's easily detachable rear engine and transmission facilitated overhauls.*

Specifications

Country: USA
Year manufactured from: 1939
Engine: four cylinder gas powered air cooled
Transmission: three-speed forward
Payload: two tons (2.03 tonnes)
Applications: city delivery
Special features: engine was detachable and mounted at the rear

Starting in 1938, White would begin production of two multistop delivery vans, dubbed Merchandor trucks. In these trucks, the engines were positioned under the floor to maximize payload. 1939 saw White build 6x6 trucks for the U.S. military. But the same year also saw White forge ahead in commercial design, with the unveiling of the White Horse.

Introduction of rear-mounted engines

The White Horse was reminiscent of the Merchandor trucks in that it was specifically designed for urban multistop delivery applications. But it differed considerably in other respects. For instance, the engine (an air-cooled Franklin-designed 40hp [29.8kw] four-cylinder, with three-speed transmission and coil rear suspension) was detachable and mounted in the rear of the truck. Otherwise, the White Horse was your ordinary everyday city truck—a straight single-bodied truck with a 2-ton (2.03 tonne) payload capacity.

The White Horse's particular combination of familiarity and innovation appealed to the commercial truck-buying public and more than 2,000 White Horses were sold in the first year they were manufactured. Production would continue until 1942, but it would grind to a halt during World War II, when White was required to prioritize the needs of the U.S. Army.

During that period, White Horse trucks were repurposed and redesignated as Model 99 and Model S-99 (with an 8ft 3in [2.51m] wheelbase), and Model 116, Model S-116, and Model H-116 (with a 9ft 8in [2.94m] wheelbase).

CHEVROLET
1940 USA

CHEVROLET KD/KE

In 1933, for the first time, Chevrolet outproduced its rival Ford. The primary division of General Motors was now able to sell a complete new truck range five years later.

Right: *Certainly one of the most attractive light trucks of its era, the 1940 Chevy had a very appealing appearance.*

Specifications

Country: USA
Year manufactured from: 1940
Engine: six cylinder gasoline powered 85hp (65kw)
Transmission: four-speed
Payload: ¾ ton (0.76 tonnes) to one ton (1.02 tonnes)
Applications: stake body, platform, cattle truck
Special features: chromed grille with "Chevrolet" letters on top

The first ¾-ton (0.76-tonne) light truck was the Model GD, created in 1937. It was named HD in 1938, JD in 1939, and KD in 1940, while the 1-ton (1.02-tonne) was called respectively GE, HE, then VA (1939) and WA (1940).

New design

The Chevy's first appearance in 1937 revealed an attractive new design. The two new trucks were very similar, but the 1-ton (1.02 tonne) vehicle was fitted with larger 14in (36cm) hydraulic brakes, bigger tires and wheels, and heavier rear spring suspension. The same year, a new six-cylinder engine was also offered. Its displacement was increased to 216.5ci (3.5 liters) and the brake horsepower was rated at 85hp (63kW). In 1939, the range received a restyled cabin, with a two-piece V windshield, and all the models had longer wheelbases. Just before the U.S. entered the war, the Chevy light looked more and more luxurious. In 1940, the distinctive radiator grille was chromed, with a big red script "Chevrolet" name on top. After this, all the trucks were fitted with new hoods, new fenders with headlights, and massive radiator grille designs.

FREIGHTLINER
1940s USA

FREIGHTLINER CF100 COE

Freightliner trucks appear to be unique in early truck history in that they saw their beginnings in the hauling industry, rather than in engine or bicycle manufacturing.

Leyland James founded Consolidated Freightlines, a hauling company and the forerunner of Freightliner, in 1929 in Portland, Oregon. In true trucker tradition, James was always looking for ways to increase profit and payload. His appetite for finding new ways to do this, particularly when it came to technological innovation, was what led to the creation of Freightliner's very first trucks.

As usual, necessity was the mother of invention. When Leyland James wanted lighter trucks that could carry more payload and stay within the 60-ft (18.3-m) length limit required by most western U.S. states, the suppliers he dealt with ignored him.

Innovations in materials and design

So, with the help of mechanics already in his employ, James started adapting trucks himself. By the mid-1930s, James and his mechanics had already fitted some of his trucks with lighter aluminum parts, including brake shoes, suspension hanger brackets, and trailer pulleys. Of course, aluminum truck and trailer bodies were the

Above: *Being in the trucking business, Leyland James of Consolidated Freightlines knew exactly what kind of truck was required so he and his team went ahead and created the first Freightliner, the CF100.*

next logical step. Next came cab-over-engine designs, because by making the tractor shorter, James could increase the length of the trailer, thus increasing payload.

The very first Freightliner cab-over-engine design to hit the road for Consolidated Freightlines was assembled from spare parts and cabs welded together by Freightlines mechanics themselves. It was pulled together from a Fageol truck with a Truck and Coach Company chassis and a custom-designed Freightliner aluminum cab. It had a Cummins diesel engine, and it was a full ton lighter than any other design with a comparable payload.

James' cab-over-engine design proved so successful that the company would build 20 more over the next three years alone. Obviously, James' idea of making trucks lighter and trailers both lighter and longer was working.

Expansion through the 1940s

By 1940, James had finally convinced other carriers, if not manufacturers, of the benefits of lighter, shorter truck designs to increase profits and payload. Consolidated Freightways (James had renamed the company) and five other western carriers formed the Freightways Manufacturing Company in 1940, to build trucks for their own fleets. The company's first manufacturing plant was opened in Salt Lake City.

James' lightweight COE design was the first truck to be produced. By this time, it had been improved on even more with a heat-treated steel alloy frame and nose for durability. Of course, the aluminum cab and trailer body would be its most distinctive and profitable feature.

But even after opening the Salt Lake City factory, James wasn't really planning to get into the manufacturing business on a permanent basis. He just wanted to build enough trucks to supply himself and the other carriers he'd gone into business with. That all changed when the Freightways Manufacturing Company decided to try to outsource the assembly of its trucks. James and his partners were quickly convinced that the products assembled outside their own plant were vastly inferior to those they'd assembled themselves. That was how Freightways stepped into the manufacturing business permanently, and how what was to become the Freightliner brand of truck was eventually born.

Specifications

Country: USA
Year manufactured from: 1940s
Engine: Cummins diesel powered
Transmission: five or 10-speed
Payload: not available
Applications: long haul
Special features: use of aluminum in parts, bodies, and cabs

Below: *James Leyland convinced five other western American carriers to form the Freightways Manufacturing Company in 1940, to build trucks for their own fleets.*

BEDFORD

1939–1945 UK

BEDFORD OY THREE-TONNER

The strategy of the British Army during World War II was to penetrate enemy lines with tanks. A wartime vehicle was needed to get both infantry and supplies to the front as quickly as possible.

Above: *The OY 4x2 Bedford military truck fulfilled a variety of roles.*

Left: *Bedford OY military 3-tonners were developed from the civilian K, M, and O models launched just at the outbreak of World War II.*

Throughout World War II, the British armed forces would depend heavily on the four-wheeled, two-wheel drive Bedford OY truck, which could haul up to 8,500lb (3,800kg) fully loaded, even though it was only designated to haul 6,600lb (3,000kg).

Wartime production

When war broke out in 1939, Bedford OY three-tonners were already being produced for commercial use. But the British War Department, which controlled some 85,000 vehicles (including more than 26,000 it "borrowed" from civilian sources), still needed more. The commercial OYs needed few adjustments to make them war ready. The manufacturer simplified the bodywork (earlier wartime bodies were made of wood) and replaced the truck's dual rear tires with single ones. More OYs would be produced during the war than any other British 3 tonners. In fact, Bedford production of OYs would equal more than one-quarter of all trucks made by Bedford during the war years (a total of 250,000 were made, OYs accounting for approximately 72,000).

OY trucks were used for a multitude of different purposes—as mobile workshops and offices, or to transport troops, machine-guns, and provisions. The Bedford OYC tanker version could carry up to 960 gallons (3,637 liters) of fuel.

Specifications

Country: UK
Year manufactured from: 1939
Engine: six cylinder gas powered
Transmission: four-speeds to bevel
Payload: three tons (3.04 tonnes)
Applications: multiple military
Special features: Armadillo had pebble-filled body walls

ALBION
1940s UK

ALBION 6X4 THREE-TON REPAIR TRUCK (MODEL FT103N)

Toward the end of World War II, many British Army trucks were in pretty poor shape. Many of them had been designed before or at the beginning of the war, and had been overhauled several times since.

Above: *The Albion FT103N repair truck was an adaptation of a 6x4 civilian model.*

Specifications

Country: UK
Year manufactured from: 1940s
Engine: four cylinder diesel
Transmission: chain or worm drive
Payload: three tons (3.04 tonnes)
Applications: military
Special features: sides of bodies could be folded down to provide work benches

The army needed a truck on the front line that could repair existing vehicles and carry spares to those who needed them. Because of a dearth of commercially available trucks, they decided to make a limited purchase of Albion 3-tonner commercial trucks as a temporary measure.

Off-road use

The Albion trucks purchased had simple commercial chassis, with six wheels and four-wheel drive. They did not have off-road suspension, but as far as the British Army was concerned, they fitted the bill.

The repair trucks carried a lathe, pillar drills, bench grinders, gas-welding equipment, battery charges, and assorted equipment. The standard Albion Clansman commercial vehicle cabs were used. The bodies had canvas tops and wooden sides, which could be folded down to provide ground-level workbenches. Canvas shelters could even be attached to the sides of the vehicle to provide more workspace on the ground. The FT repair truck was 20ft 9in (6.32m) long, 7ft 6in (2.28m) wide, and 11ft (3.35m) high with an 11ft 9in (3.58m) wheelbase.

OM

1939 Italy

OM TAURUS

Founded in 1899, the Officine Meccaniche company, in Brescia, Italy, began to build trucks in 1925. The Taurus was the most popular prewar OM vehicle, launched in the late 1930s.

Above: *The Taurus was one of the most popular prewar OM vehicles.*

The Italian company acquired the Saurer license before the thirties and built the Swiss trucks under its name. In 1933, the company came under the control of Fiat but the OM trucks were still labeled "OM."

New truck

The 1CRD was one of the most modern Italian trucks of the time. The 2.95-ton (3-tonne) vehicle was offered with a 60hp (45kw) Saurer diesel engine, a five-speed gearbox, and hydraulic brakes. Just before the onset of World War II, the "Taurus" appeared. This 3.1-ton (3.14-tonne) truck was powered by a 4.3 x 4.3in (110 x 110mm) bore and stroke, 324.65-ci (5.3-liter) diesel engine. A bus variant was launched too, as well as the "Ursus," a 6.4-ton (6.5-tonne) version powered with a bigger six-inline diesel. During the war, the Italian forces in North Africa used many Taurus vehicles. The vehicles had to cope with the appalling climate of the Western Desert, where overheating and sand-clogged filters were problems for all sides. After the conflict, a new "Taurus" went into production, but it was a completely new truck, with a cab-over-engine and a fresh style.

Specifications

Country: Italy
Year manufactured from: 1939
Engine: four cylinder diesel 75hp (56kw)
Transmission: five speed
Payload: three tons (3.04 tonnes)
Applications: platform, stake body
Special features: four inline under Saurer-licensed diesel engine

AUSTIN

1939 UK

AUSTIN K2Y

The Austin K2Y was the ambulance version of the Austin K2 ⅔-ton (0.67-tonne) truck. More than 13,000 were built for the Allied forces during World War II and most of them were still in use through the 1950s.

Above: *The Austin K2 was the main British ambulance built between 1939 and 1945.*

Specifications

Country: UK
Year manufactured from: 1939
Engine: six cylinder gasoline powered 63hp (47kw)
Transmission: four-speed
Payload: two tons (2.03 tonnes)
Applications: ambulance
Special features: ambulance body with four stretchers

Described as "Truck, 2-ton, 4x2, ambulance" in the British military list, the Austin K2Y was the most common Heavy Ambulance used by all British services. Some examples were supplied to the U.S. Air Force under "reverse lend-lease." The Russian Red Army used some K2Ys too, but it seems these ambulances were actually recaptured models—the Wehrmacht captured some trucks during the 1940 French campaign, and used them on the Russian front, where they fell into the hands of Russian troops.

Stretcher bearer

The open cab had accommodation for the driver and a medic, and the chassis was fitted with a wood frame, which was covered with a waterproof and insulated leather cloth. This body was ordered by the Royal Army Medical Corps just before the war and manufactured by the bodybuilder Mann-Egerton of Norwich, England. Four stretchers or eight sitting patients could be placed inside. There was a communicating door between the cab and the body, and a double swing door at the rear. For comfort, a heater was installed, as well as ventilators on the body sides, and two extractor ventilators in the roof. The exhaust pipe was fitted high along the body to keep the casualties away from the toxic fumes. The ambulance, fondly nicknamed "Katy," was powered by a gasoline engine with a four-speed gearbox.

PACIFIC CAR AND FOUNDRY
1942 USA

PACIFIC CAR AND FOUNDRY M26 COE

Today PACCAR is best known for having sold various heavy-duty trucks under various famous brand names, including Kenworth, Peterbilt, DAF, Leyland, and Foden.

Specifications

Country: USA
Year manufactured from: 1942
Engine: diesel powered
Transmission: four-speed forward transmission with three speed central transfer case
Payload: 12 tons (12.2 tonnes)
Applications: military tank hauling
Special features: three axles

Above: *The Pacific Car and Foundry Company of Renton, Washington, were responsible for building the U.S. Army's most powerful tractor, the Pacific M26, used to haul tank transporters.*

Pacific Car and Foundry wouldn't build a truck until 1942, when it was commissioned to do so for wartime use by the American armed forces. The U.S. Army needed a vehicle that would be powerful enough to haul tanks loaded onto semitrailers. And Pacific Car and Foundry would deliver just that, when the M26 rolled off its production line.

The M26 would be one of the most powerful trucks built during the war. It was a 6x6, in a cab-over-engine truck-tractor design that could pull 12 tons (12.2 tonnes). It had three axles and a Hall-Scott diesel engine that could develop 240hp (178.9kw). It used a four-speed transmission with a three-speed central transfer case that connected to the rear eight-wheel tandem drive as well as the front two-wheel steering axle (by shaft).

Made for tank and trailer recovery

The truck also had a winch mounted at the front and controlled from the cab and capable of pulling 35,000lb (15,876kg), so it could recover any tractor or trailer lodged in rough terrain. Two more winches were mounted at the rear of the cab.

DODGE

1942–1945 CANADA

DODGE T SERIES MILITARY TRUCKS

Most of Dodge's Canadian operations were turned over to military production in 1940. Between 1943 and 1945 the company would build a total of one million military trucks.

Right: *The wartime Canadian Dodge T series was built in a variety of forms, all of which were powered by Ford's 6-cylinder side valve petrol engine.*

Specifications

Country: Canada
Year manufactured from: 1942
Engine: liquid cooled straight six cylinder SV
Transmission: four-speeds to bevel
Payload: variable
Applications: military
Special features: rigid axles with semielliptic leaf springs

There were several T series trucks—the ½-ton 4x4 T207-WC3, the T207-WC3, the T207-WC9, the ¾-ton 4x4 T214-WC52, the T214-WC55, the T214-WC56, the T214-WC57, the T214-WC60, the T223-WC62, and T223-WC63, to name just a few. WC stood for weapons carrier but that's not all the trucks did. Some of them, but not all, had winches that could pull up to 2.5 tons (2.54 tonnes) as well.

Some—the WC52 for instance—were open-cab pickups, while others, like the WC53, were carry-alls without one. (An estimated 8,420 of the latter would be built; some of which would be fitted out with rear side doors for use as command vehicles.) The WC54 was used as an ambulance and so was built without a winch (about 29,932 were built between 1942 and 1944) while the WC55 was a gun hauler with a winch (5,380 were built during the war).

Powered by Ford motors

All WC units were powered by Ford's own T214 92hp (68.8kw), 3,200rpm, liquid-cooled straight six-cylinder SV engines with four-speed forward and one-speed reverse transmission. They had a single-speed transfer box with front axle disconnect. The clutch was dry plate.

FARGO

1940s CANADA

FARGO POWER-WAGON T-137, MODEL FM6-W300M

By the 1940s, Chrysler had stopped marketing the Fargo in the U.S., but the trucks were still widely available in Canada and abroad. In Canada, medium- and heavy-duty Fargos were manufactured domestically.

Specifications

Country: Canada

Year manufactured from: 1940s

Engine: De Soto six cylinder

Transmission: four-speeds to bevel

Payload: up to four tons (4 tonnes)

Applications: multiple

Special features: manufactured by Chrysler

Above: *Canadian-built Fargos were sold alongside other Chrysler makes including Dodge.*

When the Canadian Fargo was introduced in 1936, it was available in ½-, 1-, 1½-, 2-, and 3-ton sizes. By 1939 a 4-ton model was added to the Canadian market, but it was built in the U.S. Trucks larger than 3 tons were exported to Canada from Detroit.

The Fargo truck brand name had originated outside Chrysler, with the Fargo Motor Car Company of Chicago, which marketed a line of Fargo trucks from 1913 through 1922. With the company long dead and gone, the Chrysler Corporation took the name in 1928, and created the Fargo Motor Corporation to build and sell commercial trucks. The newly formed Fargo Motor Corporation immediately started producing two full lines of commercial vehicles: a lighter truck called the "Packet," and a larger truck, built on the Chrysler 65 chassis. Both used a mixture of Plymouth, DeSoto (owned by Chrysler), and Chrysler parts. In 1930, the company replaced the Plymouth four-cylinder engine of the Packet with a DeSoto six-cylinder engine. A 1-ton "Freighter" line was also introduced. These models also used parts from a variety of Chrysler vehicles.

Fargo emblem

Eventually, the Fargo ornament (a globe) would be attached to a wide range of vehicles, from light express and delivery vehicles to heavy dump trucks and semis. Domestic U.S. production of Fargos stopped at the end of the 1930s, after reaching 7,680 vehicles since their introduction in 1928.

OSHKOS

1940s USA

OSHKOSH W SERIES

Unlike many other truck makers, Oshkosh would survive World War I and the Depression years, even though profits would lag due to the flooding of the market with wartime trucks.

Above: *The Model W-2200 appeared in 1947. The design continued the Oshkosh tradition of snowplows—it could be fitted with plows or hydraulic fenders for ease of single-man operation.*

Right: *Oshkosh would continue to produce the W series throughout World War II and make several improvements along the way.*

Specifications

Country: USA
Year manufactured from: 1940s
Engine: six cylinder gas powered
Transmission: five-speeds to worm
Payload: not available
Applications: snow clearing
Special features: new cab design with V windshield

Models introduced through the 1920s and 1930s included trucks that would serve a multitude of applications, including fire trucks, snowplows, construction trucks, earth movers, and dump trucks. One of the company's publicity stunts during the early 1920s involved driving the trucks up local high school and library steps—a stunt that was actually quite commonly used by truck makers of that time. The first Oshkosh diesel engine appeared in the Oshkosh G series trucks in 1935. It was a Cummins.

Oshkosh started to manufacture its W series of trucks in 1940, when World War II was already underway. New with the series was an updated cab design with a V windshield. The first W series trucks were powered by 112hp (83.5kw) six-cylinder Hercules RXVC gasoline engines. These trucks would first be used by the U.S. Army to clear snow off runways. Attached to them were rotary snowblowers powered by six-cylinder gasoline engines mounted at the back of the truck.

Postwar W series production

Oshkosh would continue to produce the W series throughout World War II and make several improvements along the way. In the last year of the war, the company would introduce the W-1600 model, a large 6x6 offroader designed to haul trailers with heavy equipment. It was ideal for oilfield use.

The Model W-2200 followed in 1947. Gasoline or diesel engines were available for this truck, from Buda or Hall-Scott. The design continued the Oshkosh tradition of snowplows—it could be fitted with plows or hydraulic fenders for ease of single-man operation.

CHEVROLET
1940s USA

CHEVROLET MILITARY TRUCKS

Between 1940 and 1945 Chevrolet's focus would be on the production of military vehicles. In fact, General Motors would produce no less than $12 billion worth of military vehicles during the war.

Above: *Chevrolet's extra-lightweight Jeep was powered by a 90-degree V-type air-cooled 20.5hp (15.2kw) Indian motor. It didn't do very well in testing and was soon dropped.*

Specifications

Country: USA
Year manufactured from: 1940
Engine: four or six cylinder gas powered
Transmission: multiple options
Payload: up to one ton (1.016 tonne)
Applications: military
Special features: cross-vehicle transmission for vehicles of more than 50 tones (50.8 tonnes)

The war gave occasion for many technological advances at Chevrolet and GM. The transmission developers would take the opportunity to experiment, designing a specialized tank transmission and steering system known as a "cross vehicle" for trucks weighing above 50 tons (50.8 tonnes). The new technology was a hit, with Chevrolet producing more than 60,000 trucks in 1943 and more than 71,000 in 1944 equipped with the new transmission and steering—more than enough proof that they worked, and under the toughest of conditions.

Failed Experiments

Some experiments, of course, would fail. One such was an extra-lightweight Jeep powered by a 90-degree V-type air-cooled 20.5hp (15.2kw) Indian motor. It had a tubular frame, four-wheel independent suspension and transverse leaf springs, inboard rear brakes, and integral transmission and transfer case for its four-wheel drive. But it didn't get very far in testing and was dropped early on. Another such vehicle was the M38 Armured Car. Produced in 1945, it was discontinued when the war ended.

Above: *Most of the Chevrolet trucks made from 1942 to 1945 were developed specifically for the military. In fact, General Motors would produce no less than $12 billion worth of military vehicles during the war.*

SCANIA
1945 Sweden

POSTWAR SCANIA L SERIES

Postwar, Scania-Vabis' truck range was very much a product of what the Swedish truckmaker had been forced to learn during the war, largely thanks to the demands of the Swedish Army.

Above: *Postwar, Scania would continue to expand its L-series range in the modular direction. Different models, such as this L21, shared the same components, including cylinder heads, pistons, connecting rods, bearings, and exhaust systems with other Scania trucks.*

Specifications

Country: Sweden
Year manufactured from: 1945
Engine: four, six, or eight cylinder gasoline or diesel powered
Transmission: four-speeds plus optional four auxiliary gears
Payload: variable
Applications: multiple
Special features: helical gearwheels

They had wanted modular trucks and that's what Scania would provide. Modular principles in development and production give the customer an almost unlimited choice to tailor a vehicle to a specific transportation task. At the same time, modularity ensures that as few components are used as possible, which is an advantage in terms of servicing and parts supply.

Truck range expands

Scania would continue to expand the modular development of its L-series range after the war, with the introduction of the L/F10 and the L/20/LS20 lines. (The L/F10 was a 4x2 and the L20/LS20s were 6x2s.) The trucks came in a choice of four wheelbases and were powered by four- or six-cylinder "Royal" engines (Scania had developed its first diesel engine in 1936, but producer gas models were also still available.)

During the late 1940s, the company made up for time lost during the war, bringing out a steady stream of technological improvements for its trucks, including hydraulic vacuum-assisted brakes, an optional auxiliary gearbox that doubled the number of gears (to eight), and helical gearwheels.

The innovations would include yet more engines, including four-, six-, and eight-cylinder models available in carburettor and diesel versions. The modular principle was again applied, with shared components including cylinder heads, pistons, connecting rods, bearings, and exhaust systems.

Through the late 1940s and early 1950s, Scania-Vabis would earn a reputation as one of Europe's foremost engine producers.

WILLYS-OVERLAND
1940s USA

WILLYS-OVERLAND FOUR-WHEEL DRIVE UTILITY TRUCK

The Standard Wheel Company of Indiana started out building bicycles. But in 1903, the company decided to expand its operation to include the Overland Automotive Division.

The new division's first vehicle would be the Overland "Runabout." John North Willys would purchase the division in 1908. Runabout production continued but it would be moved to a new manufacturing plant in Toledo, Ohio. The Willys-Overland Company would be founded in 1912, and added to its production range with the Willys-Knight series automobile and the popular "Whippet." The company would suffer during the Depression, however, and go bankrupt in 1936, necessitating a reorganization of the company as Willys-Overland Motors Inc.

By 1940, the newly revived Willys-Overland Motors would start making trucks

Above: *The Willys-Overland Motors company would start making trucks for the American military in 1940.*

for the American military. The company's first venture would be a prototype for the American army's first four-wheel drive ¼-ton utility truck.

Willys-Overland won the military contract after a competitive bidding process and began production of the vehicles (called "Jeeps") in 1941. The company produced roughly 350,000 of these trucks throughout the 1940s. Postwar, the four-wheel drive trucks were used for a variety of applications and Willys-Overland continued to develop its commercial truck line as well as an international distribution network.

Multiple chassis uses

By the time the Korean War broke out, Willys-Overland had designed the M38 utility truck and its later version the M38A1, both of them improvements on the original 4x4. The company also developed the M170 frontline ambulances, which extended the body and chassis of the M38A1 to accommodate patients on stretchers. More than 150,000 M170s would be produced.

In 1953, Willys-Overland were purchased by the Henry J. Kaiser and renamed Willys Motors Inc. But the company would keep manufacturing military and nonmilitary trucks for the U.S. government. The next truck was an M274 weapons carrier, and in 1960 the company won its bid to produce a ¼-ton delivery van for the U.S. Post Office.

Inset: *The Jeep became the most widely known and respected four-wheel-drive vehicle on the market during this period.*

Below: *The Jeep name arose from "G P" meaning "General Purpose."*

Specifications

Country: USA

Year manufactured from: 1940

Engine: four cylinder gasoline

Transmission: three- or four-speed

Payload: not available

Applications: military

Special features: four-wheel drive

GMC
1940s USA

GMC CCKW SERIES AND OTHER WARTIME TRUCKS

The GMC CCKW series of military trucks were made between 1941 and 1945, specifically for the U.S. military during World War II. No fewer than 600,000 Model 353s and 25,000 Model 352s were built in just four years.

Above: *If one truck stuck in the minds of World War II veterans it was the ubiquitous "Jimmy" 2.5-ton (2.54-tonne) 6x6. The GMC CCKW and its 6x4 sister the CCW were built in vast numbers.*

They came in several different shapes and sizes, for example, the CCKW-353 had a 13ft 8in (4.1m) wheelbase and was a 6x6, while the CCKW-352 was a 6x4 with a 12ft 1in (3.7m) wheelbase. Both trucks weighed about 5,512lb (2,500kg). Both trucks had six-cylinder overhead-valve diesel engines that could develop up to 90hp (67.1kw) and a maximum speed of 44.7mph (72km/h). The 353 and the 352 both had five-speed overdrive-style gearboxes with two-speed transfer cases, and could go about about 16 miles to the gallon (400km on 182 liters of fuel). Both trucks were available with hard or soft cabs, and with or without winches. But the U.S. wasn't the only country producing GMC trucks.

Germany takes over

When World War II began in Europe, Germany nationalized the GMC Opel Division. The plant immediately started producing trucks for the Axis powers. By then GM had been established in Europe for some time. In 1923, GM's first European assembly plant was opened in Copenhagen under the name General Motors International A/S. It was to build Chevrolets for sale in Scandinavian countries, the Baltics, Germany, Poland, Czechoslovakia, Austria, Hungary, and Russia. The first GM vehicle to come off the Copenhagen assembly line at the beginning of 1924 was also the first GM to be built outside North America—it was a Chevrolet utility truck. GM went on to establish operations in Antwerp, in Belgium, in 1924.

Vauxhall Motors Ltd., in Luton, England, was acquired by General Motors in 1925. The same year saw General Motors Export Division open its first warehouse in Malaga, Spain, as well as a sales office in Paris and a warehouse in Le Havre, France. Also in 1925, GM opened a sales office in Berlin.

The year 1927 would also see a GM assembly plant built in Berlin. By 1929, GM had acquired Adam Opel AG, based in Rüsselsheim, Germany. Opel would be taken over by the German government in 1940. You might say GM took the war personally.

Dedicated to the war effort

The company converted 100 percent of its production to the war effort as of 1942. During World War II, GM delivered more than $12,300,000,000 worth of war material to lead the Allied war effort, including aircraft engines, aircraft parts, trucks, tanks, marine diesels, guns, shells, and miscellaneous products.

In 1942 alone, the company would produce 148,111 power units solely for military use, which amounted to 16.9 percent of the total truck production in the U.S. that year. By 1943, the company would produce no fewer than 16 different trucks for the U.S. Army. They included 1½- to 3-ton small-arms repair trucks, 1½-ton radio transceiver trucks, 1½- to 3-ton air-compressor trucks, 2½-ton 6x6 artillery tractors, 2½-ton 900-gallon (3,400-liter) tankers, 1½- to 3-ton machine-shop trucks, 2½-ton cargo and troop transporters, 2½-ton cavalry haulers, 1½- to 3-ton earth-boring trucks, 1½-ton recruiting trailer trucks, 2½-ton searchlight carriers, 8-ton arsenal tractor trailers, COE navy fuel tankers, 4-ton anti-aircraft defense trucks, and 1½-ton 4x4 troop transporters.

GMC also produced an estimated 560,000 2½-ton 6x6 trucks for general use. But perhaps the most distinctive vehicle built by GMC during World War II was the amphibious DUKW, know as the "Duck." The vehicle looked like a boat with six wheels (duals in back) and a soft top. It was designed to carry up to 50 men on either land or water.

Inset: *It was basic in design but the famous "Jimmy" was the mainstay of the Allied Forces in World War II.*

Specifications

Country: USA
Year manufactured from: 1941
Engine: six cylinder overhead valve gas and diesel
Transmission: five-speed overdrive with two-speed transfer case
Payload: variable
Applications: military
Special features: available with hard or soft cabs

Below: *The amphibious variant of the CCKW, designated DUKW, would traverse land and water equally and could cope with 50 fully equipped soldiers. The vehicles were nicknamed "Ducks."*

FEDERAL

1941 USA

FEDERAL 6X6 WRECKER

MANY OF THE BIGGEST ARMY TRUCKS EVER USED BY UNCLE SAM ARE NOW BEING BUILT BY FEDERAL

FEDERAL TRUCKS

Above: *Though building for both civilian and military purposes, Federal concentrated on military production for World War II. This patriotic advertisement makes the point.*

From 1941, Federal trucks would focus on military vehicle production, while continuing to manufacture for the commercial market. In fact, Federal started producing 6x6 wreckers as soon as World War II began.

Left: *Federal wreckers were powered by Hercules six-cylindered engines.*

Federal introduced the truck, with a payload capacity of 7.5 tons (7.62 tonnes) in 1941. The design was also produced by REO. The wreckers had Hercules six-cylinder engines that could develop up to 180hp (132.4kw). They also had five-speed transmission and came in two ranges. Air springs were used in front, which was unusual for trucks at the time.

Wartime production

Wartime production at Federal also included a 6x4 heavy-duty truck tractor with a 20-ton (20.3-tonne) payload capacity and a Cummins six-cylinder diesel engine that could develop up to 130hp (96.9kw).

All the while Federal made sure to continue advertising its trucks to the general public, touting its dedication to wartime production for the military as proof of the durability and reliability of its vehicles.

The advertising campaign and the truckmaker's dedication to military truck production eventually paid off. In 1944, the U.S. government, which was regulating wartime vehicle production because of building supply shortages, allowed Federal to produce 2,034 trucks for civilian use. Federal would thereby be in a much better position than most other truckmakers, both in the U.S. and abroad, at the end of World War II: By the end of the war, the average American-owned truck was 7½ years old.

Postwar the company prospered, having just filled an order for the U.S. military worth $15 million at the beginning of 1945. Prewar styling would dominate the first Federals to be introduced after the war, which would keep tooling costs down.

Specifications

Country: USA
Year manufactured from: 1941
Engine: six cylinder two stroke diesel powered
Transmission: chain or worm drive, five-speeds
Payload: not applicable
Applications: wrecker
Special features: air springs on front axles

INTERNATIONAL

⚒ **1941 USA**

INTERNATIONAL H-542-11 COE

Through the 1930s, International's range had expanded to include commercial vehicles offering payload capacities ranging from 1.5- to 10-ton (1.52- to 10.16-tonne) six-wheelers.

Above: *This truck had a distinctive slanted hood and a bull nose.*

Specifications

Country: USA

Year manufactured from: 1941

Engine: four to six cylinder diesel powered

Transmission: five-speeds to bevel

Payload: up to five tons (5.08 tonnes)

Applications: military

Special features: cab-over-engine design, slanted hood and bull nose

International trucks were powered by everything from a four-cylinder Waukesha diesel engines to six-cylinder engines built by the truckmaker itself.

Innovations in design

International had also innovated in terms of design—the company had produced its first cab-over-engine model in 20 years in 1935 (the C-300 COE).

Other design changes followed and proved popular, such as the redesigned cabs, pontoon fenders, split windshields, and passenger car looks of the D series models. (The sales of D-2 models were considerable—no fewer than 80,000 units were sold in the three years from 1937 to 1940.)

By 1940, the K series would replace the D series and sleeper cabs would already have been introduced. In that year alone, International built 86,000 power units. The company was already selling more 2-ton (2.03 tonne) trucks than any other truckmaker in the United States and would be among the top three U.S. truckmakers in terms of production volume.

chapter 5

The Postwar Era

Two world wars in 20 years didn't just decimate the ranks of truckmakers on a global scale, it also led to the creation of a few new companies along the way. Some, such as PACCAR, even saw their beginnings thanks to one or another of the wars, mostly due to the need for trucks to serve the military. PACCAR, among others, would keep their focus on trucks and abandon their previous manufacturing businesses.

Above: *The 1950s International Harvester was ideally suited to a wide range of applications.*

Left: *The Peterbilt 351 was developed in response to a growing Interstate highway network, which was making long-distance freight haulage more practical than ever.*

Technology would evolve further in the wake of World War II, with truckmakers once more forced to develop commercial vehicles that were less expensive to run and easier to maintain. Several companies also found that diesel engines suited the needs of their customers better, if they hadn't already made the switch from gasoline to diesel prior to the war. Diesel was (and remains to this day) substantially less expensive than gasoline. The use of aluminum in truck and trailer construction also grew more common, with lighter trailers and trucks allowing for more payload, which was important, because the distances trucks were traveling to pick up loads were much longer. This itself was beause continent-spanning road systems were fast being developed.

Trucks that traveled longer distances needed to provide overnight accommodation for their drivers. That meant the development of cabs with bunks, refrigerators, and storage in them, and in-cab heaters would be needed in colder climates. It didn't take long for consumers to get used to what trucks brought, especially as trucks were also being used to advertise their wares, like rolling billboards extolling the merits of everything from soda to construction supplies. The arrival of the consumer age would herald the first driver shortages.

FORD
1946 UK

FORD COE TRUCK

By the end of World War II, truck- and carmakers were being allowed to manufacture freely again. So automotive companies had to retool their plants for mainly civilian production.

Ford was in a fair financial position to do so, given that the company had not only produced military vehicles but also had been allowed to continue civilian production of public transit vehicles throughout the war. Among the vehicles produced for this purpose by Ford were the Model 29B and Model 49B transit buses. Also introduced by Ford to the civilian market during the war were two new trucks: the Model 498T and the Model 494T. Both trucks had 100hp (74.5kw) V-8 engines.

By the end of 1945 Ford was in the enviable position of being able to offer 42 different truck chassis models. Model 59 V-8 proprietary engines powered all of them at first, although Ford would offer trucks with six-cylinder engines later the same year. Some trucks in the prewar lineup were not reoffered in 1945, however, despite their popularity leading up to the war. These included the COE trucks, which Ford had introduced in 1939.

Ford reoffers COE

While 1945 would be a very bad year for some truckmakers in the United States it would be excellent (in relative terms) for Ford—the company would produce 122,473 civilian and military trucks during the year. The excellent sales figures would pay off so much that in 1946 Ford was in a position to reoffer its platform, COE, and sedan delivery trucks with the same new design and engineering as the trucks it had introduced in 1945.

Above: *Starting in 1938 Ford began offering "COE" versions of their light to medium trucks.*

Above, top: *Ford styling reflected the tastes of the new postwar era.*

Specifications

Country: UK
Year manufactured from: 1946
Engine: V-8, four or six cylinder diesel powered
Transmission: four-speed to bevel
Payload: variable
Applications: multiple
Special features: cab-over-engine semi-forward design

AUTOCAR
1946 USA

AUTOCAR

Autocar played an important role in World War II by producing trucks for use by the military. However, after the war ended the company quickly returned to building civilian trucks.

Below: *Autocar trucks were popular among owners thanks to their workhorse nature, but equally popular with drivers because of their comfortable cabs.*

Right: *Autocar's two-pieced, slightly curved windshield would be copied by other manufacturers in years to come.*

Specifications

Country: USA
Year manufactured from: 1946
Engine: full range of White gasoline and Cummins diesel engines
Transmission: not available
Payload: not available
Applications: long-distance, on-highway, logging, off-highway
Special features: pioneered the use of alumnum in cab construction, shaving about 25 percent off weight.

It wasted no time jumping back into the fray of the domestic trucking industry, building more than 5,000 Autocar trucks in 1946 alone.

Popular Class 8 provider

Autocar was particularly focused on the Class 8 market and owners and drivers alike were fond of the look and feel of the company's latest Class 8 offerings. It was involved in different applications at the time, depending on the region. In the east it was popular for moving heavy equipment, while in the west Autocar trucks were used to haul everything from logs to livestock. In 1950, Autocar developed a new cab. This was a big hit within the industry and would later go on to find a home on White and Diamond Reo trucks. It featured a two-piece, slightly curved windshield.

Despite its impressive sales figures, Autocar was mired in financial troubles. This led to its acquisition by White Motor Company in 1953 at which time production was relocated to Exton, Pennsylvania. It wasn't long after the acquisition that Autocar became one of White's premier truck offerings. The trucks were marketed to niche applications such as the logging, mining, quarrying, and oil and gas sectors of the trucking industry. After the acquisition Autocar was given a new-look grille to reflect its new identity.

VOLVO
🛠 **1946 Sweden**

VOLVO LV15/LV24

While Volvo always seemed to be a leader when it came to cutting-edge technology in trucks, the manufacturer was one of the last truckmakers to make the switch to diesel-powered engines.

Above: *Volvo was among the last truck manufacturers to switch to diesel-powered engines. The first Volvo truck to make the switch was the LV15.*

Above: *Although the LV15 didn't boast the highest horsepower, it was still used to take on the toughest tasks, such as log hauling.*

Specifications

Country: Sweden
Year manufactured from: 1946
Engine: Volvo's first generation of diesel engines
Transmission: five-speed to bevel
Payload: not available
Applications: long-distance, on-highway, off-highway, vocational
Special features: Volvo finally adopted diesel engines with this lineup.

Even so, by 1940 Volvo was ready to make the switch to diesel engines, but when war broke out the company put its plans on hold. It was 1946 before Volvo launched its precombustion type diesel engine. The first truck to be equipped with Volvo's initial generation of diesel engines was the LV15. It was very similar in appearance to the "Roundnose" trucks of 1939/1940, but with a larger hood to house the big diesel engine.

Progressively more powerful

Volvo's original diesels produced only 95hp (70.8kw), but that gradually increased and had already hit 100hp (74.5kw) by 1949. Despite the relatively low output, the LV15 and its successor, the LV24 model, were capable of handling very rigorous transportation duties. These trucks were commonly used in the construction industry and to transport goods over long distances.

In 1950, with Leylands help, Volvo took its diesel engine technology to the next level by introducing a direct-injection VDC engine, which delivered much better fuel mileage. Today's diesel truck engines are based on this concept.

INTERNATIONAL
1947 USA

INTERNATIONAL KB

In 1947 International continued its K series of trucks with the introduction of the KB. The new KB had a wide grille, which extended out on the front fenders.

Right: *International's KB series of trucks had a full range of sizes, including this pickup truck. Despite their small size, they were used for a wide range of hauling tasks. Before long, trucks of a similar design were surfacing in other parts of the world.*

Specifications

Country: USA
Year manufactured from: 1947
Engine: up to an International 360ci (5.9- liter), 126hp (193.9kw) Red Diamond
Transmission: five-speed to bevel
Payload: 1 ton (1.016 tonne)
Applications: local pickup and delivery.
Special features: this series only enjoyed two years of production, yet was a highly popular truck during that postwar era.

The KB series didn't enjoy a particularly long production span, only being built between 1947 and 1949. During this time, production numbers were quite high, however. For instance about 64,000 KB-5s alone were constructed during this period. This was largely attributable to the postwar demand for trucks.

Panel bodies

The smaller models had panel bodies and could be used to haul modest loads. Wooden boxes could be found on the back of each of the models allowing for freight haulage. These trucks were commonly equipped with dump bodies. Tractor versions were also available for hauling semitrailers.

It was a stylish-looking truck with a sleek front-end and large round headlights. A long conventional-style hood provided the truck with a big-truck look and feel. It also carried the relatively new International logo featuring a red "I" over a black "H." Today, the KB series of trucks are popular choices for truck restorers. The largest of the KB series, the KB-8 had an International 360ci (5.9-liter), 126hp (93.9kw) Red Diamond engine under its hood.

Typical specs on the KB-1 included a heavy-duty clutch, rear heavy-duty springs, West Coast mirrors, and split-rim wheels with eight-ply tires. The popularity of the KB series of trucks was noticed overseas as well as in North America.

FREIGHTLINER
1947 USA

FREIGHTLINER MODEL 800 "BUBBLENOSE"

Freightliner introduced its all-aluminum "Bubblenose" Model 800 cab-over-engine in 1947. The company was prospering in the postwar years and it was aggressively pursuing the use of aluminum in its truck parts.

Specifications

Country: USA
Year manufactured from: 1947
Engine: Cummins diesel
Transmission: nine-speeds to bevel
Payload: not available
Applications: long-distance, on-highway
Special features: one of the first mainstream trucks to fully utilize aluminum in an effort to minimize weight and maximize payloads.

Above: *Freightliner's Bubblenose may have looked a bit unusual, but its aluminum composition allowed carriers to increase payload by about 0.98 ton (1 tonne).*

Aluminum alloys were being widely used in Freightliner's cabs, axle housings, brake drums, cross members, wheels, radiators, and even frame rails. The company was adamant about developing a lightweight vehicle that didn't compromise strength or safety. By doing so, Freightliner trucks were able to haul 0.98 ton (1 tonne) more payload than most other trucks on the market.

The company had plenty of incentive to maximize payload because the first Freightliners were built by and for Consolidated Freightways. Since Freightliner's birth it had been providing Consolidated Freightways with reliable transportation. However, engineers had now built a truck that was deemed ready to compete on the open market. The very first sale, to Vince Graziano, a produce hauler based in Portland, Oregon, came in 1948. He chose the Model 800 all-aluminum Bubblenose and he was so impressed by the truck that he ordered five more shortly thereafter.

In 1949 Freightliner offered a sleeper-cab version of its trucks. Later, in 1951 the company put White Motor Company in charge of sales, service, and distribution of Freightliners.

GM

1947 USA

GM ADVANCED DESIGN TRUCK

GM's first Advanced Design trucks, introduced in 1947, may have been considered light-duty to most, but they were certainly capable of hauling respectable loads for a truck of their size.

Right: *General Motors postwar GMC truck range benefited from bold new cab styling.*

Specifications

Country: USA

Year manufactured from: 1947

Engine: GM 90hp (67 kw) six cylinder

Transmission: four-speeds to bevel

Payload: 0.98 ton (up to one tonne)

Applications: local pickup and delivery, farm

Special features: a spacious, comfortable cab and improved visibility were priorities addressed in this line of GM trucks.

What immediately struck the public was the completely revised design and appearance of the new trucks. This was based in part on discussions GM held with business owners. What the company found was that the primary concern for owners and operators was a spacious cab providing comfortable seats and improved visibility. GM complied with the Advanced Design truck, adding 8in (20.3cm) of width and 7in (17.7cm) of length to the cab.

The increased cab size gave the new generation of GM trucks a much larger appearance than their predecessors. In addition, larger windows improved the vehicle's safety and visibility. GM trucks were often used for transportation, but at that time they were mainly a work truck used in many different applications. Drivers who would spend much of their workday in the truck were concerned about the vehicle's comfort and to this end GM introduced a fresh-air heater/defroster that would draw fresh air into the cab from outside while forcing old, stale air out through vents at the cab rear.

Further improving safety, the Advanced Design truck featured a fully welded cab, which provided additional strength.

SCANIA

1947 **Sweden**

SCANIA LS23

As a neutral country, Sweden wasn't directly affected by World War II, so civil truck production continued as before, although the fuel shortages were acute. Postwar, Scania launched a new range: the L20 series.

Above: *A two-color Scania L20 series in action. The cab seems old fashioned, but it was spacious and comfortable.*

In fact, a new series had been built in 1944–the F10 series (with a four-inline diesel engine) and the L10 (with a four-inline gasoline engine). This range was completed by a heavy-duty truck series: the L20 (4x2 axle configuration) and the LS 20 (6x2 axle configuration), and the LS 23, a 6x2 configuration truck strengthened for a more than 9.84-ton (10-tonne) payload. The L20 was a 10.8-ton (11-tonne) truck and the LS 20 a 14.7-ton (15-tonne) truck.

Powerful engine

Four wheelbases were available and a big reliable diesel engine powered the L20 range. Called the "Royal," the 518.70ci (8.5-liter) engine was a serious commercial threat to the other Swedish manufacturer, Volvo.

In 1949, a direct injection was fitted to replace the precombustion chamber system. This revolutionary design significantly improved the fuel economy. In addition, the driver's job was made substantially easier by an optional auxiliary gearbox that doubled the number of gears to eight, and there were helical gearwheels as well. The hydraulic vacuum-assisted brakes were a very important improvement, too.

Massive but modern, the big LS 20 and LS 23 series were built only for two years. The heavy 6x2 remained rare because of its high price and also because of competition from the railroads. Only 280 of them were manufactured.

Specifications

Country: Sweden

Year manufactured from: 1947

Engine: six cylinder diesel engine, 115hp (85.8kw)

Transmission: four-speed (or eight with optional auxiliary gearbox)

Payload: 10 tons (10.1 tonnes)

Applications: lumber

Special features: diesel direct-injection engine (1949), hydraulic vacuum air-assisted brakes

COMMER
1948 UK

COMMER QX

In 1948 Commercial Cars Limited (Commer) of the United Kingdom unveiled the Commer QX, featuring an underfloor engine and plenty of chrome. The chrome gave it a sharp appearance popular with haulers.

Specifications

Country: UK
Year manufactured from: 1948
Engine: six cylinder
Transmission: five-speed
Payload: 12 tons (up to 12 tonnes)
Applications: fire, long distance, on-highway, military
Special features: Commer's underfloor engine was copied throughout the industry.

Above: *The Commer QX, introduced in 1948, featured an underfloor "slant six" gasoline engine. Starting in 1954 Commer offered their unique TS3 two-stroke diesel alternative.*

The Commer QX was available in a wide range of sizes and configurations. By 1951 the truck was available as an Express 8cwt van as well as a 12 ton (12-tonne) highway tractor. Several other options in between those two extremes were also available. It also commonly served as a fire truck.

Colorful history

Commer got its start in London in 1905 and played an important role in the World War I. About 3,000 Commer military vehicles were built by 1919, and they enjoyed a reputation for being tough and dependable. After the war, however, the company struggled to find its place in the commercial market. Eventually, Humber took over production in 1926 and it was later swallowed up by Rootes Group. Commer was among the first British truckmakers to offer the Perkins diesel engine.

The Commer QX's underfloor engine paved the way for other manufacturers. The similar Karrier models also adopted this style soon thereafter. In addition to being a popular choice for fire apparatus, the Commer QX also saw some military duty.

DAIMLER-BENZ

⚒ **1948 Germany**

THE ORIGINAL UNIMOG

Following World War II, former Daimler-Benz engineer Albert Friedrich began developing an agricultural vehicle that would combine the characteristics of a farm tractor with those of a heavy-duty truck.

Friedrich's original plan was to develop an all-wheel drive vehicle with four wheels of the same size that would be capable of speeds of up to 31mph (50km/h) in top gear.

Intended for agricultural use, the Unimog was first revealed to the public in 1948. The first mass-produced Unimogs were built in Boehringer, Germany, where production continued until 1950, at which time Daimler-Benz took over the company and moved production to Gaggenau. Prior to the takeover, 600 Unimogs had been built in Boehringer. While Daimler-Benz took over production of the Unimog in 1950, the brand's original ox-head logo was not replaced by the Mercedes star until 1956.

Covering new ground

While the Unimog was proving itself as a valuable agricultural vehicle, observers began to take notice of a wide range of other applications the unique vehicle would be well suited for, with some subtle enhancements. In 1953, the 401/402 series was launched, which featured an enclosed roof. The power and diversity of the Unimog also soon caught the interest of the military and in 1955 Daimler-Benz introduced the

Above: *The original Unimog was designed for agricultural purposes, but users soon realized that its capabilities were virtually endless. The truck combined the characteristics of a farm tractor with those of a heavy-duty truck.*

longer-wheelbase Unimog S, which was designed for military service. Construction of the Unimog S continued until 1980 and it established itself as an indispensable vehicle within the German, French, and American militaries, among others. The Unimog S was originally powered by a gasoline-driven car engine that only delivered 82hp (61.1kw), but in 1956 the 401/402 Unimog was given a more powerful engine in 1956 (an increase of 30hp [22.3kw] and renamed the Unimog 411. A longer-wheelbase version was also released and soon after the vehicle was available with a synchronized gearbox.

Unimog's first diesel engine

In order to keep up with the increasing number of roles that the Unimog was being called on to meet, Daimler-Benz introduced the medium-heavy series 406 in 1963 with a wheelbase of 93.7in (238cm). Finally, at this time, a diesel engine was matched with the vehicle, increasing its power output to 65hp (48.5kw). Because of its new higher horsepower rating, the latest generation Unimog was named the U 65. The renaming of the Unimog to reflect its horsepower continued with later versions being called the U 80, U 84, U 100, U 110, etc. In order to fill the void between the original Unimog and its vastly more powerful cousins, Daimler-Benz introduced the lightweight 421/403 series.

Milestones

In May 1966, the 100,000th Unimog rolled off the assembly line. Ironically, however, very few of these vehicles were actually used for agricultural purposes, as was the original intention. So Daimler-Benz introduced a similar vehicle custom-built for agricultural applications, called the MB-trac. This was designed with Unimog engineering and the brand continued until 1991.

In 1974, the Unimog received yet another more powerful engine. The U 120 was part of the 425 series and featured a new, modern-looking cab and larger hood. It was the birth of the design the Unimog has carried relatively unchanged to this day. The 125hp (93.2kw) version of the Unimog boasted a permissible gross weight of up to an amazing 8.8 tons (9 tonnes).

Trendsetter

The Unimog was a trendsetter when it came to new technology. A case in point is the use of disk brakes, which were fitted on the Unimog long before becoming standard on European highway trucks.

Daimler-Benz continued releasing many updated models of the Unimog throughout the 1970s, essentially providing a version that would be ideal for each customer's individual needs. In 1977, the company celebrated the production of the 200,000th Unimog—another milestone.

Inset: *The Unimog turned a lot of heads when first introduced. Observers didn't know whether the unusual machine was a tractor or a heavy-duty truck.*

Specifications

Country: Germany

Year manufactured from: 1948

Engine: four cylinder diesel

Transmission: five-speed

Payload: 2.5 tons (2.54 tonnes)

Applications: agricultural, military, off-highway

Special features: originally developed as an agricultural vehicle, the Unimog found a number of other niche markets where it became popular.

Below: *The Unimog was ahead of its time. It featured disk brakes a long time before they became standard equipment on European highway tractors.*

AEC

1948 UK

AEC

Associated Equipment Company (AEC) became known as Associated Commercial Vehicles (ACV) in 1948 after taking over Maudslay Motor Company and Crossley Motors.

Specifications

Country: UK
Year manufactured from: 1948
Engine: AEC six cylinder
Transmission: not available
Payload: not available
Applications: not available
Special features: not available

Meanwhile, the AEC range was expanded with the introduction of the medium-duty Mercury model in 1953, which was driven by the AV4160 six-cylinder diesel engine. Later, this engine was further developed and eventually became known as the more powerful AV470.

AEC produced vertical (AV) and horizontal (AH) versions of its six-cylinder in-line engines and there were many variants of both types with different power outputs. For a short time the company would also begin producing a series of V8 engines.

New trucks introduced

The Mercury was also further developed and it gave birth to the twin-steer Mustang in 1956 as well as the six-wheel Marshal, which was introduced in 1960.

Associated Equipment Company was also building heavier-duty trucks including the Mk. IIIs, which included the Mandator 4x2, the Mammoth Major 6 6x2 and 6x4, and the Mammoth Major 8 8x2 and 8x4. The Mandator was also available in a tractor-unit version. Most of the trucks were offered in a hooded version, generally for export.

Above: *AEC was a leading UK manufacturer of heavy trucks like this 1949 "Mammoth Major" Mk. III eight wheel rigid.*

FORD

1948 USA

FORD F SERIES

Ford premiered its F series range of trucks in 1948, which included the heavy-duty F-3 to F-8 trucks. This line of trucks would go on to become one of the best-selling lines in automotive history.

Above: *Ford's F-series with their bold, modern styling were best-sellers in the 1940s and 1950s.*

Above, top: *The F-series was the first of the Ford conventionals.*

Specifications

Country: USA

Year manufactured from: 1948

Engine: range of Ford engines

Transmission: four-speeds to bevel

Payload: Up to 21,500 lbs (9,752kg)

Applications: Local pickup and delivery

Special features: One of the best-selling lines in automotive history.

Production of the F series began at Ford plants in Norfolk, Virginia, and Atlanta, Georgia, but 14 other plants were soon also participating. In 1948 alone, 290,000 F series vehicles were built, which marked the highest volume for Ford since 1929. In the postwar era, many customers needed dependable work trucks for agricultural, delivery, or towing applications.

The 1948 F series trucks had a tough image thanks in part to broad fenders and a horizontal grille, which combined to give it a strong, bulky look. The F series' headlights were mounted directly into the fender rather than being perched on top, giving it a sleeker look. Ford made no bones about the tough looks of its F series. In promotional material the company proudly stated that "Appearance-wise, Fords are the handsomest, truck-iest-looking trucks on the road." Ford also assured customers these trucks would last 10 years or more. The styling of the F series truck line remained relatively unchanged through 1952.

Full range

The lightest-duty F series was the F-1 with a mere half-ton capacity. Still, it was considered a workhorse of a truck and was promoted as the "Master of 1,001 light-delivery jobs." The heaviest truck in the line was the F-8, which boasted a carrying capacity of 21,500lb (9,752kg). Ford continues to produce the F series today.

DAF
⚒ **1949 Netherlands**

DAF A-30, A-50, A-60

In 1928, Hubert van Doorne started what would later become the DAF vehicle company with his brother, Wim, in Eindhoven, in the southern part of the Netherlands.

By 1932, the brothers had founded the Van Doorne Trailer Company. (DAF is the abbreviation for Van Doorne's Automobilefabrik in Dutch.) The company would at first include just four employees and would be located in a small workshop in the corner of a brewery. DAF would start out doing welding, engineering, and forging work for canal boats in Eindhoven, but a year later, there would already be 32 employees on the payroll.

The Great Depression in the late 1920s and early 1930s would force the company to diversify, starting to build trailers and semitrailers. One of the company's first inventions would be the DAF demount trailer, specially developed for loading and unloading railroad containers for intermodal rail and road transportation. DAF would become one of the first suppliers of container trailers in the world.

A-30 production begins

In 1936 the brothers would start making military vehicles for the Dutch army, which led to the production of large commercial trucks. It also led to the creation of the "TRADO-drive" product, which was a conversion kit that would allow the Dutch armed forces, which were low on funds, to purchase new four-wheel drive vehicles, to convert existing commercial two-wheel drive vehicles to four-wheel drive for military use.

DAF Trucks started commercial production

Above: *This DAF A60 was built after 1956, as indicated by the grille. Originally, the front grille had seven horizontal bars across it. The number was reduced to six in 1956.*

in Eindhoven in 1949 with the model A-30. The first A-30s (built in small numbers at first—the factory hadn't yet managed to attain the efficiencies of mass production) were assembled from parts purchased from other manufacturers. The trucks were powered by Hercules or Perkins diesel engines, rode on axles from Timken and switched gears with gearboxes provided by Eaton.

A series expands

The A-30 was a 3-ton (3-tonne) truck and it was joined on production lines by a number of other trucks. The A-50 was a 5-ton (5-tonne) offering and DAF also introduced an A-60 6-ton (6-tonne) truck. In 1950 DAF introduced the A-10 0.98-ton (1-tonne) van, which formed the base for the A107 light-duty pickup truck, developed for export purposes. DAF soon developed special chassis for refuse and dumper applications and also began producing military trucks.

Cab and chassis construction

DAF itself focused entirely on the construction of the chassis and cabs and relied on other manufacturers' proven components. The exceptions to this rule included brake components and wheel hubs, which were manufactured by DAF in-house.

The A-30 was a cab-over-engine truck at a time when they were still relatively new to the market, giving the truck a modern look compared to traditional conventional-style trucks. The grille was immediately recognizable by the seven horizontal chrome strips. In 1956, the number of strips was reduced to six.

By 1955, DAF had produced its 10,000th truck chassis. More expansion was immediately required as the company had just received an order for more than 3,500 more vehicles. Also in that year, the Hercules gasoline engines were discontinued in favor of the new Perkins R6 diesel or the Leyland 0.350 engines. DAF eventually acquired the licensing rights to the Leyland 0.350 and began producing that engine itself in a new engine factory. This was initially done to guarantee the quality of the engines, but it wasn't long before DAF was producing its own versions of the engine. DAF was one of the first truck manufacturers to make use turbocharged engines in its vehicles.

DAF continued expanding by beginning to build its own rear axles in 1957. This is the same year the company began using ZF gearboxes—DAF still felt it would be too costly to build transmissions in-house.

Inset: *DAF built this experimental prototype in 1948.*

Specifications

Country: Netherlands

Year manufactured from: 1949

Engine: Leyland or Perkins diesel engines

Transmission: five, six and eight speeds to bevels

Payload: 6 tons (up to six tonnes)

Applications: local pickup and delivery, long-distance, on-highway, military

Special features: led to development of the "TRADO-drive" product, a conversion kit allowing the Dutch armed forces to convert existing commercial two-wheel drives to four-wheel drives for military use.

Below: *DAF's A series of trucks featured a "racey" grille consisting of two female skaters.*

SOMUA

1946 France

SOMUA JL 12

The *Societe d'Outillage Mécanique et d'usinage d'Artillerie* (Mechanical Tools & Artillery Foundry Company) was founded during World War I. Its high-quality trucks entered the market only during the prewar period.

Somua was famous for two automotive outputs: fire trucks and fire apparatus, and military vehicles, especially tanks. After World War II, Somua launched a new range of buses and a new truck. The truck was a modern and fast 6.8-ton (7-tonne) payload forward-control truck powered by a Hesselman licensed engine. This engine (used at first by Volvo) could run on gasoline, gas, or alcohol with few adjustments. According to the mixture, the four-cylinder engine offered from 70 to 105hp (52 to 78kw). But this engine was irreconcilable with the new French postwar planning regulations and Somua was ordered to join with two other manufacturers, Willeme and Panhard, in order to standardize the production. The JL 12 changed its Hesselman engine for a more classical Panhard diesel engine. This six-cylinder powerplant was at the time one of the most advanced diesel engines in the world. With the Panhard engine, the JL 12 became finally the JL 15. In 1946, only three JL 12s were built, but in 1949 197 chassis rolled off the production lines. The truck was fitted with air-assisted hydraulic brakes and the multileaf springs were very effective. The much-appreciated comfortable wide cab was given an improved range of front visibility. The JL 19, launched in 1955, was a redesigned version with a new facelift and a bigger 150hp (111kw) engine.

Above: *The Somua JL 12 was maybe too modern and too expensive to have any commercial success.*

Specifications

Country: France

Year manufactured from: 1946

Engine: four cylinder Hesselman engine, 70 to 105hp (52 to 78 kw)

Transmission: five-speed

Payload: 6.8 tons (seven tonnes)

Applications: platform, stake body, covered truck

Special features: four inline under Hesselman-licensed diesel engine

VOLVO

1953 Sweden

VOLVO VIKING L38

In the early 1950s Volvo unveiled what would become one of its most famous trucks ever—the Viking. It seemed a fitting name for the model since both Volvo itself and the ancient Vikings originated in Scandinavia.

Above: *The Viking was one of Volvo's most famous trucks ever.*
Above, top: *The Viking utilized a diesel engine that was known for its reliability.*

Specifications

Country: Sweden
Year manufactured from: 1953
Engine: Volvo 424.9ci (7 liter) up to 100hp (74.5kw).
Transmission: not available
Payload: not available
Applications: long-distance, on-highway, logging, off-highway
Special features: the Viking established itself as one of Volvo's most famous trucks to ever take the highway.

The Viking was based on Volvo's L24 Roundnose truck. When first introduced, the Viking featured only a new hood and fenders. However, the manufacturer continued to enhance the Viking throughout the 1950s and into the early 1960s. Volvo continuously updated not only the engine strength of the Viking but also its chassis. Power steering was another feature that wasn't available on the earliest models.

International appeal

The Viking gained worldwide appeal due largely to its reliability, but it was not an overly powerful truck. In the 1950s, enormously powerful engines weren't required for road transportation in most of Europe since heavy loads were not yet the norm. The Viking was powered by a modest nominal 424.9ci (7-liter) engine, which was known for its dependability, if not its power. The original Viking engine had an output of just 100hp (74.5kw).

Throughout the 1950s, the engine received power upgrades to keep up with trends in the road transportation industry. In 1954, a genuine 424.9ci (7-liter) engine was introduced (the initial one was actually just over 364.2ci [6 liters]) along with wider fenders resembling the ones that could be found on Volvo's L39 Titan—the Viking's bigger, more powerful cousin.

The majority of Vikings manufactured were two-axle versions, however a three-axle variant was available as well. Production of the Viking ended in 1965 when it was replaced by the similar-looking N86.

INTERNATIONAL HARVESTER

1950s USA

INTERNATIONAL HARVESTER

In the 1950s, International Harvester's L and R series trucks were being used in a variety of applications. Many owners claimed the trucks were run by the most reliable big gasoline truck engines in the U.S.

Specifications

Country: USA

Year manufactured from: 1950s

Engine: full range of International big-block gas engines

Transmission: not available

Payload: not available

Applications: long-distance, on-highway, fire, logging, off-highway

Special features: engine reliability made this truck famous and it's still a firm favorite among truck historians.

The International Harvester could be customized in a variety of configurations with a number of different engine choices, depending on the size of the truck. A typical medium-duty Harvester could be paired with a six-cylinder engine, but that eventually gave way to a small block V8.

The heavy-duty Harvesters were powered by big-block gasoline engines at this time. In the 1960s International Harvester produced a gasoline-powered V8 549 engine, but the trucking industry was already embracing diesel engines and it failed to take off.

Similar appearance

It took a trained eye to differentiate between the R and L series of International Harvester by appearance alone. Both versions of the truck were available in a variety of configurations including both dual and single drive.

Frequently, operators would remove the side skirting from the hood to provide better cooling, especially while driving at slow speeds during the summer. Occasionally, removing only the right-hand side skirting was sufficient as that side of the truck housed the manifold and carburettor.

Above: *The International Harvester was ideally suited to a wide range of applications including log hauling. The truck's owners sometimes felt the Harvester was invincible and they were commonly overloaded.*

BEDFORD
1950s UK

BEDFORD

After the war Bedford resumed production of its trucks using the K, M, and O models introduced in 1939 as its base. However, it wasn't long before the company replaced those trucks with more current offerings.

Above: *This S type Bedford was dubbed "The Big Bedford."*

Specifications

Country: UK
Year manufactured from: 1950s
Engine: full range of Bedford gasoline engines
Transmission: not available
Payload: 6.8 tons (7 tonnes)
Applications: long-distance, on-highway, military, local pickup and delivery.
Special features: Bedford produced full range of commercial vehicles following World War II.

The S type was known as "The Big Bedford" and it vaulted the manufacturer into the 7-ton (7-tonne) class. The truck also featured a new forward-control design. The S type was powered by a 300ci (4.9-liter) gasoline engine. It was replaced by the low-cab forward-control TK in 1960. Later is was offered with Leyland and Perkins diesels.

R type embraced by the military

The four-wheel drive R type derivative was designed in 1950 and was embraced by the military, where it served as its standard 3.9-ton (4-tonne) truck until being replaced by the MK and MJ types, and then hit the civilian market in 1953. Meanwhile, Bedford had produced the CA van, which went on to become one of its bestsellers.

In 1953 Bedford continued its aggressive product launch with the introduction of the A type, which was followed by the D and J types. These models were built well into the 1970s, with the exception of the A type, which was terminated in 1957.

The J type (or the TJ) was launched in 1958—the same year Bedford produced its millionth truck. It featured a single-piece windshield and 16in (40.6cm) road wheels providing a low body platform height on models up to 3.9 tons (4 tonnes). The TJ had a bonneted cab.

PETERBILT

1950 USA

PETERBILT 280/350

Peterbilt founder T.A. Peterman passed away in 1945 and it was a tragedy that he never lived to see one of the most innovative of the truck company's products—the Model 280/350 "Bubblenose."

This truck was developed in response to a growing Interstate highway network, which was making long-distance freight haulage by truck more practical than ever. Because not all American truckers were fond of the cab-over-engine appearance, the company also released the 281/351 conventional-styled trucks. As a result, Peterbilt had its bases covered when it came to the on-highway market.

The Bubblenose

Things were looking good for Peterbilt in 1950. The company had reached sales of

Above: *The Pete 350 "Bubblenose" was a cross between a conventional and cab-over design. Most American truckers preferred a conventional truck design, but this style incorporated the best of both worlds.*

$7 million and the trucking industry was becoming increasingly enamored with the truck. Still, in the rapidly changing transportation industry no company could relax. They had to continue exploring new technologies and treading new territory. Thus, the 280/350 "Bubblenose" was born.

The 280/350 model was referred to as the "Bubblenose" because it did not have a flat front like other cab-over-engines. Instead, the engine protruded slightly from the front of the cab, giving the truck what looked like a small, rounded nose. However, that nose was far shorter than the traditional conventional-style hoods of the day and the truck was classified as a cab-over-engine. Some truckers speculated the Bubblenose resulted from a shortening of the 1949 Model 360, which had a short nose that resembled a half-hood conventional.

COEs gain acceptance

The Bubblenose's cab was perched quite high compared to other models on the road at the time. The engine could be accessed by tilting the cab, but this wasn't easy to accomplish according to some truckers. They often chose to access the engine and other components by swinging the front fenders out of the way.

Two years of road testing went into the 280/350 before it was mass produced for sale to the public. The Bubblenose was a popular truck because it adhered to length restrictions being imposed in some states, especially in the eastern USA.

The Bubblenose was replaced by the 352 in 1955. The 352 was a redesigned cab-over featuring a slightly-curved four-piece windshield and it remained in Pete's product line until 1982 when the Model 362 superseded it. The 362 would have a more modern one- or two-piece flat windshield.

The 281 and 351 conventionals

Introduced in 1954, the 281 and 351 conventional models gave customers an alternative to the Bubblenose. These trucks were well liked by drivers and owners and remained in production for 11 years. They were among the first to be adorned with Peterbilt's signature red oval logo. Ironically, cab-over-engine versions of these models were introduced in late 1955.

Specifications

Country: USA

Year manufactured from: 1950

Engine: Cummins or GM

Transmission: 10-speed

Payload: not available

Applications: long-distance, on-highway

Special features: these trucks helped pioneer the Drom-Box, which were cargo boxes attached to the back of the tractor.

Left: *The 351 was the all-American classic "conventional" tandem drive tractor, seen here with a typical sleeper compartment mounted behind the cab.*

DIAMOND T

1950s USA

DIAMOND T 950-951

Diamond T had made a name for itself during the war by producing what were widely thought to be the best-looking trucks in which to do battle. However, the postwar years would not be so kind to the company.

One postwar innovation was the development of a "comfort cab," which was built in conjunction with International Harvester. The 1950 cab featured a number of enhancements, such as detachable fenders and a hood that allowed outside air to pass into the engine compartment to provide better cooling and improved fuel economy. The cab also featured a four-point "diamond" mounting of the cab, sheet metal, and radiator, and had a curved windshield and concave instrument panel.

To prove how well the cab was constructed, the company subjected it to hours of severe twisting at the Chicago Auto Show that year, but no structural damage took place.

Diamond T 950-951 introduced

That same year, Diamond T introduced its 950 and 951 series of trucks, which were built at the company's Chicago factory. The 950 was built with Western truckers in mind, because they tended to demand more horsepower and larger radiators to provide adequate cooling for the larger engines. The 950 and 951 Diamond Ts were so big that, although they were a conventional design,

Above: *The Diamond T 921-C tilt cab trucks of the mid-1950s could gross up to 38 tons (34.5 tonnes) and had a choice of Cummins diesels.*

the driver was often perched as high as he would be in a cab-over truck.

Diesel takes over

The first of these trucks to hit the streets were equipped with gasoline engines. However, those soon gave way to the more efficient diesel-powered engines. In 1953 these trucks were paired with the tilt-cab.

In the 1950s Diamond T was focusing exclusively on the heavy-duty truck market and was only building trucks with 3 tons capacity or greater. The 950 and 951 were the largest offerings and were typically powered by a 300hp (223.7kw) Cummins diesel engine or a 280hp (208.7kw) Buda diesel.

Diamond T also built a handful of 950 RS trucks, which may well be the largest Class 8 trucks ever built for on-highway applications. It has been reported that fewer than 80 of these trucks were ever built. They were powered by Cummins diesel engines.

Diamond T honoured

In 1953 Diamond T's tilt-cab cab-over-engine truck was presented with the prestigious National Design Award for an industrial project. It was the first time any truck design had won this coveted prize. The technology which earned Diamond T the award was a counterbalance system, which no longer made it necessary for a power unit to tilt the model 723C cab. This development was adopted by other truckmakers in years to come.

The last preacquisition Diamond Ts

While the trucking industry was awash with rumors regarding the impending acquisition of Diamond T, the company continued releasing new models. During the 1950s the company built 6x6 trucks for the military, which were powered by 224hp (167kw) Continental R6602 engines. The trucks featured automatic six-wheel drive and power steering.

The year before Diamond T was acquired by White Motor Company, it introduced the Model 831 which was driven by a 239hp (178.2kw) Hall-Scott 590 engine. This engine could burn liquid petroleum gas, which boosted its power to 256hp (190.8kw). A tilt-cab model 911C diesel was also offered before Diamond T was taken over.

Diamond T was experiencing trouble by the late 1950s. Its competitor Reo was also struggling to compete with the big players such as International, GM, and White. White Motor Company stepped in and purchased both companies in 1958 and consolidated the two brands.

Left: *The tilt-cab design of the Diamond T won it the coveted National Design Award in 1953. It marked the first time a truck design had ever received the award.*

Specifications

Country: USA
Year manufactured from: 1950s
Engine: 300hp (223.7kw) Cummins diesel engine or a 280hp (208.7kw) Buda diesel
Transmission: 10-speeds to bevel
Payload: not available
Applications: long-distance, on-highway, construction, off-highway
Special features: Diamond T gained fame during World War II and was known as the "best looking trucks to do war."

Below: *The tilt cab fitted to Diamond T medium-weight trucks from the early 1950s had a modern appearance and was imitated by other manufacturers in subsequent years.*

VOLVO
1951 Sweden

VOLVO TITAN

One of the most famous trucks ever to hit the highway was Volvo's Titan, introduced in 1951 for long-distance haulage and rigorous construction-site applications.

Above: *The Titan featured an engine built from lightweight components. It made it an expensive option but helped customers maximize payloads.*

Legal wrangling had prevented Volvo from using the name "Titan" in its initial sales and marketing efforts, but the problems were eventually resolved and the truck was officially christened the Titan.

Rounded design

The Titan incorporated many of Volvo's proven components, but fitted to a new-look truck that featured a rounded design similar to those trucks that could be found in North America around this time.

The first Titan was essentially a redesigned L29 C diesel longnose, but it featured a much-improved interior. As a result, driver comfort was greatly enhanced, and a more efficient engine provided for an overall more efficient truck. The engine was a direct-injection 582.7ci (9.6-liter) VDF, compared to its VDB pre-combustion chamber predecessor which only had a capacity of 552ci (8.6 liters).

Turbocharged engine

The VDF engine in the original Titan was constructed out of lightweight materials typically used in the construction of aircraft engines. While the engine's lightweight design may have offered some benefits when it came to efficiency, it was an expensive engine to build. Therefore, Volvo resorted to a more conventional D96 engine made of cast iron a short time later.

One of the Titan's features that helped contribute to its fame was the turbocharged engine it was given in 1954. The Titan became one of the first production trucks in the world to be equipped with a turbocharged engine, although they had been previously used in ships, locomotives, and aircraft.

Given the success of turbocharged engines in other modes of transportation, Volvo developers John Stålblad and Bertil Häggh set to work creating a smaller turbocharger which would fit under the hood of the Titan.

Remarkable results

While the turbocharger increased the vehicle's curb weight by 55lb (25kg), the results were worth the small sacrifice in weight. The engine output increased by a remarkable 35hp (26kw) from its original 150hp to 185hp (111.8kw to 137.9kw).

While the turbocharged engine contributed greatly to the success and storied tradition of the Titan, it wasn't the only feature that caused this vehicle to stand out. The L39 and L49 Titan was also one of the first to offer many other technological advances that were just now finding a home in commercial vehicles. Air brakes replaced traditional hydraulic braking systems for instance, and power steering reduced driver fatigue and allowed for better handling. Volvo's safety cab, introduced in 1959, was also a feature on the Titan. This cab provided drivers with a safe environment in which to work and was developed with the help of the Swedish national safety administration.

The Titan was a well used heavy-duty truck known for its power. Production of the vehicle continued until 1964.

A new Titan—the TIPTOP F88

In 1964 Volvo introduced the Titan TIPTOP F88, which featured a tilt cab that made servicing the engine and other components much simpler and less time-consuming. At the time, the Titan's main rivals didn't offer this convenience and so the new Titan was immediately welcomed by the transportation industry. The sleeper was also improved to increase driver comfort, particularly for long-haul truckers.

Another reason for the F88COE's success was the new engine, transmission, and chassis components, which were derived from the Volvo N88. The engine itself was a 583ci (9.6-liter) powerplant specially prepared for turbocharging, making it capable of high power outputs. An eight-speed synchronized transmission made driving easier and more efficient.

In 1970, the Titan TIPTOP G88 was introduced to comply with legislation that required a certain distance between the front and rear axle on a vehicle combination.

Left: *In addition to running long-haul, the Volvo Titan was commonly used for rigorous construction applications as well.*

Specifications

Country: Sweden
Year manufactured from: 1951
Engine: Volvo VDF direct-injection 582.7ci (9.6 liter)
Transmission: eight-speed synchro to bevel
Payload: not available
Applications: long-distance, on-highway, logging, off-highway
Special features: lightweight engine was built from materials traditionally used to build airline engines.

Left: *You would expect to see a conventional-style truck like this in North America, not Sweden. Volvo, however, decided to give the longnose design a shot in its home country.*

KENWORTH

⚒ **1951 USA**

KENWORTH 853 AND CAB-BESIDE-ENGINE

In the 1950s, Kenworth had established itself as a major force in the trucking industry. In fact, by 1952 the manufacturer's trucks were hauling 16 percent of all land freight.

Above: *The Kenworth 853 was famous for its heavy-duty applications. It was crucial in developing the Middle East oil reserves.*

Specifications

Country: USA

Year manufactured from: 1951

Engine: full range

Transmission: full range

Payload: not available

Applications: long-distance, on-highway, off-highway, oil drill rigs

Special features: truck was used to help develop oil reserves in the Middle East.

Kenworth's 853 was designed in 1951 for the Arabian oil company ARAMCO. About 1,700 units were sold to the company, making it a hugely successful launch. The truck played a major role in the development of Middle East oil reserves.

Cab-beside-engine

In 1954 Kenworth shocked the trucking industry when it unveiled its new cab-beside-engine (CBE) truck. Again, the company was hoping to achieve unprecedented efficiencies and it was able to provide an additional half ton of cargo as a result of the change.

Other benefits included improved vision for the driver, which was especially welcomed by truckers who frequented mountain roads. The driver and passenger would be seated one in front of the other in the CBE. The exhaust pipe was located on the hood, as was the right-side mirror. A bunk could be added at the back of the cab.

However, the truck never did really take off. The passenger seat was small and the truck lacked interior space.

SCANIA

1954 Sweden

SCANIA-VABIS REGENT

Scania-Vabis released its hugely successful Regent in 1954. The truck became Scania's flagship model and continued to carry this distinction for many years afterward. The Regent was nicknamed "King of the Road."

Above: *The Regent helped Scania-Vabis to dominate the heavy-haulage market in the late 1950s.*

Specifications

Country: Sweden
Year manufactured from: 1954
Engine: not available
Transmission: five and 10-speeds to bevel
Payload: not available
Applications: long-distance, on-highway, forestry, construction, off-highway
Special features: the Regent made Scania a true player in the heavy-haulage market.

The Regent took on demanding duties such as long-distance haulage, construction, and forestry. It was generally powered by a six-cylinder direct-injection diesel engine capable of 150hp (111.8kw).

Paving the way for growth

During the year of its debut, Scania-Vabis sold 1,800 trucks in Sweden alone, making it a true competitor to the likes ofVolvo. It was considered a huge success for the truckmaker, which until the late 1940s didn't even have an adequate sales and service organization. Prior to the war, the majority of Scania's vehicles were sold to companies that had their own repair facilities. Many of these vehicles were buses that were operated by municipalities. After the war, however, Scania-Vabis realized it had to develop a dealer network that could also service the vehicles. Ultimately, the manufacturer teamed up with Volkswagen and Willys-Overland, importing these vehicles along with its own trucks and helping launch a successful dealer network.

This growth allowed Scania-Vabis to make inroads in the heavy-haulage sector of the trucking industry. Its market share in the late 1950s was often between 40 and 50 percent in Sweden.

MACK
1953 USA

MACK G, H AND B SERIES

The 1950s were busy years for Mack, which introduced a number of advancements to its product line. The G, H, and B series trucks were among the most significant developments from the company.

In 1953 Mack introduced its new Thermodyne open chamber, direct-injection diesel engine, which set a new standard in performance and fuel efficiency. Mack's decision to make its own engines available on its trucks was seen as a bold move at the time. Other engines, including those produced by Cummins, could also be specified.

H series

The H series was introduced in 1953 and was dubbed the "Cherry Picker" because of its extremely high cab (8ft 10in [2.7m]). This

Above: *Nicknamed the "cherry picker," this truck featured an elevated driver's seat. The short BBC measurement allowed it to haul 35-ft (10.6-m) trailers.*

truck had a very short BBC, which allowed it to haul 35-ft (10.6-m) trailers. At the time the maximum overall legal length limit in many regions was 45ft (13.7m), making this a practical configuration.

They were typically powered by Mack's ENDT 673 Thermodyne, which produced up to 211hp (157.3kw) at 2,100rpm. A Quadruplex 18-speed, two-stick gearbox was another common spec. The Cherry Pickers were replaced by Mack's popular Ultra-Liner.

B series

The B series, also introduced in 1953, became one of Mack's most successful products to hit the highway. It was an easily identifiable truck thanks to its stylish, rounded appearance. The cab resembled that of a 1940s-style pickup truck rather than the big, boxy highway tractor cabs typical of this era.

More than 127,000 of these trucks were sold before production was discontinued in 1966 in favor of the R series. This line of trucks was known for its longevity and many of them are still in active service.

G series

Mack began production of the G series in 1959 and produced a total of 2,181 units by 1962. The truck featured an all-aluminum cab to reduce its weight and to allow for the haulage of heavier loads.

Left: *The stylish, rounded appearance of Mack's B series made it an instant hit with North American customers. The truck was a refreshing change from the "boxy" truck designs that dominated the highway at the time.*

Specifications

Country: USA

Year manufactured from: 1953

Engine: Thermodyne open chamber, direct-injection diesel or Cummins

Transmission: Quadruplex 18-speed, two stick

Payload: not available

Applications: long-distance, on-highway, severe-service

Special features: H series "cherry picker" provided high sitting position for improved view of the road.

Below: *The G series featured an aluminum cab, which made it possible to haul heavier loads. It was especially popular in the West Coast region of North America.*

GUY

1954 UK

GUY INVINCIBLE RANGE

In 1954, Guy Motors, a manufacturer based in Wolverhampton, England, stepped up its heavy-duty truck program and expanded the weight range of its line of trucks.

Above: *The now-defunct Guy line of trucks boasted impressive payload capabilities for its day. It could haul up to 31.4 tons (32 tonnes) with a trailer or 23.5 tons (24 tonnes) without.*

Specifications

Country: UK
Year manufactured from: 1954
Engine: full range of Gardner, Meadows, Leyland, AEC, Cummins, and Rolls Royce engines
Transmission: not available
Payload: GVW up to 31.4 tons (32 tonnes)
Applications: long-distance, on-highway, vocational, off-highway
Special features: originally named the Goliath line, this series was renamed due to a legal conflict with an existing van line.

The new heavier-duty trucks were available in 4x2, 6x2, 6x4, 8x2, and 8x4 configurations. The company referred to the trucks as the "Goliath" line at first, before being told that that name had already been claimed by a German van manufacturer. So in the end, Guy named the new heavy-duty trucks the "Invincible" range.

The Invincible trucks combined an AEC chassis design with diesel engines provided by either Gardner or Meadows. The heaviest truck from this line had a gross vehicle weight of 31.4 tons (32 tonnes) with a trailer.

Meet the Warrior

In 1955 Guy introduced the 13.7-ton (14-tonne) gross vehicle weight Warrior, which featured a Motor Panels cab. An even more powerful version was released and it was dubbed the Formidable. The latter was primarily available as a tractor unit.

The Invincible was replaced in 1958 with an even more heavy-duty line—the Invincible Mk. II. This latest addition to the Guy family had a completely redesigned chassis that was built to be extra durable. A wide variety of engines could be specified.

ROTINOFF

1955 UK

ROTINOFF ATLANTIC

British manufacturer Rotinoff Motors Ltd. was founded by and named after White Russian immigrant George Rotinoff. He built the company in Colnbrook, near Slough, in Buckinghamshire, in 1952.

Above: *The Rotinoff Atlantic was built for military applications, such as tank transportation.*

Specifications

Country: UK

Year manufactured from: 1955

Engine: Rolls Royce C6.SFL series 109 direct-injection engine

Transmission: 12-speed

Payload: GCW up to 137.7 tons (140 tonnes)

Applications: severe-service, on- and off-highway, military.

Special features: the Super Atlantic boasted GCWs twice the amount of the original.

Some of those trucks ended up taking on civilian transportation duties, but most were used by the military. In 1955 the company introduced the Rotinoff Atlantic GR.7, a 6x4 tractor that had undergone successful trials with the Swiss armed forces.

Massive GTW

The truck had a maximum gross train weight of 137.7 tons (140 tonnes). The Atlantic GR.7 was powered by a 743ci (12.17-liter) Rolls Royce C6.SFL series 109 direct-injection engine. The diesel powerplant was capable of up to 250hp (186.4kw). The truck was also equipped with a 12-speed transmission and Kirkstall axles.

Eventually an even more powerful engine was available for the Atlantic GR.7. This was a C6.TFL Rolls Royce turbocharged diesel with 275hp (205kw) capabilities.

Super Atlantic produced

As if that wasn't enough power, Rotinoff added an even heavier-duty Super Atlantic GR.7, which was driven by the 988ci (16.2-liter) "straight eight" Rolls Royce powerplant which had a power output of up to 335hp (249.8kw).

These units featured 15- or 18-speed gearboxes and had gross train limits of up to 295.2 tons (300 tonnes)—more than twice that of the original Atlantic!

DAIMLER-BENZ

1955 Germany

DAIMLER-BENZ LP SERIES

Daimler-Benz introduced its LP series of cab-over-engine trucks in 1955 when the LP 315 7.7-ton (7-tonne) truck took to the roads. The "P" stood for "Pullman" in reference to the comfort of a Pullman car's cab.

Above: *The LP series of Mercedes-Benz trucks had a cab that was intended to reflect the interior of a luxury car. To reiterate this point, the "P" in LP stood for "Pullman," a popular luxury car of that era.*

The new trucks lacked the classic fenders and cab shape that previous generation of L series trucks were known for, and instead it had a rounded, comfortable cab design with an interior likened to that of a luxury car.

The dawn of the cab-over-engine

Prior to the introduction of the LP series, most of Daimler-Benz's trucks were hooded. COE trucks, however, were beginning to find their place in the market. The LP 315 was joined in 1958 by the LP 333 twin-steer six-wheel truck which had a 15.7-ton (16-tonne) gross vehicle weight or 31.5-tons (32-tonne) gross train weight. It was the first year the German government had given the green light to such heavy loads.

The LPS333 is added

An LPS333 tractor was also introduced to the market and it was also rated at 31.5 tons (32 tonnes) gross combination weight. This truck's second steering axle was mid-mounted 5ft 3in (1.6m) ahead of the drive axle. Each of the LP trucks featured a full-width cab and a split windshield. However the LP 321, LP 322 and LP 337 would later be available with a smaller cab and single windshield. These later additions to the LP family were offered with payloads of 5.9 to 6.8 tons (6 to 7 tonnes).

Easier to understand design codes

By 1963 Daimler-Benz's cab-over-engines had undergone many changes. The LP 1620 was introduced with a gross weight of 15.7 tons (16 tonnes) and a 200hp (149kw) engine. The model names, which were obscure in the 1950s to say the least, were now easier to understand. The LP 1620, for example, referred to its gross weight and the engine horsepower.

Customers were impressed with the cab's simple, cubic design. Extended versions of most models were released with the option of a sleeper compartment for long-distance hauls. The new generation of LP trucks were generally driven by Daimler-Benz's proven six-cylinder in-line engines featuring an efficient direct-injection system and power ratings in the 210hp (156.5kw) range. It wasn't long before three-axle trucks were added to the product line.

Lightweight LPs

A new lightweight series of LP trucks was also introduced consisting of trucks with weights starting at 5.9 tons (6 tonnes) (LP 608). These cab-over-engines eventually replaced the short-nose trucks Daimler-Benz had been producing. This line of lightweight COEs was designed to look much like its bigger cousins with an angular cab, rectangular grille, and attractive glazing. The light LP cabs were firmly mounted to the frame and were not tiltable.

More powerful engines demanded

In the mid-1960s Daimler-Benz launched the LPS 2020 three-axle rear-steer tractor unit, which was engineered to capitalize on an increase in allowable weights from 31.4 tons to 37.4 tons (32 tonnes to 38 tonnes). Because of the heavier loads anticipated as a result of the regulatory change, more powerful engines were required.

Daimler-Benz developed a new series of engines to meet these new demands and offered them in V6, V8, V10, and V12 sizes. Their power capabilities ranged from 192hp to 400hp (143.1kw to 298.2kw).

Some of the first trucks to be matched with engines from the new family were the LP and LPS 1632, 2032, and 2232 models. These trucks were identifiable because of their deep front wheel arches. The first of the new engines to hit the road was the OM403 which was a 976ci (16-liter) V10.

As a result of all these new vehicle introductions, Daimler-Benz had developed a broader product line than any other commercial manufacturer of the day. The company was a player in nearly all significant sectors of the trucking industry.

Specifications

Country: Germany

Year manufactured from: 1955

Engine: Daimler-Benz six-cylinder in-line engines

Transmission: four-speeds to bevel

Payload: not available

Applications: long-distance, on-highway, local pickup and delivery.

Special features: LP line of trucks catered to virtually every application and GVW range.

Below: *With the German government allowing heavier loads than ever, the LP was put to work delivering a range of heavy freight.*

HUMBER
1955 UK

HUMBER PIG MILITARY VEHICLE

The 0.98-ton (1-tonne) Humber Pig military vehicle earned its nickname because of the way its round headlights and snoutlike grille gave the vehicle a piglike appearance.

This 4x4 armored truck was developed for the British military in the 1950s. It was originally introduced as a general-purpose truck.

Two-man crew

The "Pig" was piloted by a two-man crew but could carry between six and eight men. The two-man crew would be situated in the middle and the troops would sit at the rear. It featured a Rolls Royce B-60 in-line six-cylinder engine under the hood with 160hp (119.3kw) and a top speed of 45mph (72.4km/h). The transmission was a five-speed manual with a reverse speed as well, much like the street vehicles found in North America at the time.

The hull of the truck was constructed of welded steel armor plate to provide as much protection as possible against enemy fire. Even the tires were extra durable, able to run flat for about 50 miles (80km) if deflated.

As terrorists began using more deadly weapons, the Royal Electrical and Mechanical Engineers increased the armor of the vehicle in Operation Bracelet. The vehicle had to then be further modified to handle the increased weight. The Humber Pig was able to take on many different tasks in a military environment.

Above: *The Humber "Pig" earned its nickname thanks to its long, square nose like the snout of a hog.*

Specifications

Country: UK
Year manufactured from: 1955
Engine: Rolls Royce B-60 in-line six cylinder
Transmission: five-speed manual transmission
Payload: 0.98 ton (one tonne) GVW
Applications: military
Special features: nickname comes from the snoutlike front end.

KENWORTH

1956 USA

KENWORTH MODEL 900 SERIES

Kenworth's reign as an independent company came to an end in 1956 when Pacific Car and Foundry stepped in. At this time the company was renamed Kenworth Motor Truck Company—a division of PACCAR.

Above: *A fiberglass tilting hood was becoming the norm on Kenworths with the 900 series of trucks. These trucks were extremely popular among owner-operators thanks to their unique styling.*

Specifications

Country: USA

Year manufactured from: 1956

Engine: Cummins or GM diesel

Transmission: full range

Payload: not available

Applications: long-distance, on-highway, vocational, off-highway, oilfields

Special features: tilt hood replaces the previously common butterfly-style hood.

However, PACCAR did not interfere with Kenworth's custom philosophy and dedication to quality. Shortly after the new marriage, the new Kenworth introduced the Model 900 series. The truck featured a new frame design with a dropped front section providing a lighter and shorter chassis. In the late 1950s Kenworth's GRP tilting hood was becoming the norm. Customers could still order the older butterfly-style hood but it was slowly being phased out by the new style. By 1965, the tilting hood would be a standard specification on all of Kenworth's on-highway trucks.

Putting the 900 to the test

The truck immediately established itself as a workhorse. A fleet of 923s were put to work in the rugged northern Yukon Valley where they helped move more than 2,952.7 tons (3,000 tonnes) of supplies and equipment over 385 miles (240km) of ice and tundra. The weather was often -60°F (-51°C), so the trucks worked around the clock and were never shut down. The trucks were powered by Cummins NH 200 diesel engines and compression brakes.

Kenworth found the 900 series trucks able to handle the cold weather with no effect on performance or frequency of engine breakdowns. This was despite the fact the ice would sometimes collapse, dropping the trucks into up to 4ft (1.2m) of ice and water. It was truly a challenge for the 900 series Kenworths. They delivered in the end, however, surpassing everyone's expectations.

DAF
1957 **Netherlands**

DAF 2000 SERIES

DAF decided to plunge into the international transportation sector in 1957 with the introduction of its 2000 series. It was a short-lived experiment, however, with the line being discontinued just two years later.

It was a unique truck, with a 9.8-ton (10-tonne) capacity rear axle—the only truck to offer this at the time. The cab was based on the A1300/1600/1800 series cab. More room was added at the front, however, to provide more cooling power.

Normal control vehicle

At about the same time the 2000 series was made available, DAF also released a normal control vehicle for specific market segments. This was known as the 12 and 15 series. In 1958, DAF began the in-house production of axles and this resulted in the need for a completely new factory. The new facility was built in Oevel, near Westerlo in Belgium. It remains in use for the production of axles today.

Above: *The 2000 model of 1957 was DAF's first truck designed for long-distance international haulage. Just five years after its launch it was superseded by the much improved 2600 model.*

The Eurotrailer unveiled

DAF continued adding new products to complement the 2000 series. One new addition to the product line was the Eurotrailer, introduced in 1962. This was a lightweight semitrailer with an integral aluminum body. DAF also surprised many during the same year by adding the 2600 truck to its lineup.

The DAF 2600 had a compact design that enabled increased payload and longer load lengths. The truck was able to accommodate a sleeper for long-distance haulage. It was quickly dubbed "The Mother of International Road Transportation."

DAF 2600 added

The 2600 launched DAF into the realms of the most respected long-distance truck manufacturers because of its increased driver comfort. It also helped DAF attain increased market share and sales and in 1964 the company produced its 50,000th truck chassis. The following year, after leading his company to success, DAF founder Hub van Doorne retired. During his reign, the company had acquired more than 100 patents.

In 1973 DAF replaced the illustrious 2600 with the 2800, which featured a cab that was 7.8in (20cm) wider. The sleeper cab could fit two full-size beds and the truck was powered by a 704ci (11.6 liter) DAF engine.

Below: *Day and sleeper versions of the old DAF cab were available. The day cab shown here is on a 4x4 model for on- or off-road construction work.*

Specifications

Country: Netherlands

Year manufactured from: 1957

Engine: not available

Transmission: six-speeds to bevel

Payload: not available

Applications: long-distance, international, on-highway

Special features: this truck was ultimately unable to make an impression in the long-distance international transportation segment.

DAF

1957 Netherlands

DAF TORPEDO

DAF started off as a manufacturer of cab-over-engine trucks, but the company eventually gave in to demands for a conventional-style cab. The result was the torpedo-front chassis introduced in 1957.

Above: *Despite the preference for COE trucks in Europe, DAF offered their "Torpedo" conventional for the needs of certain customers.*

Inset: *The DAF Torpedo was available in a range of body styles that enabled it to take on a variety of hauling duties.*

Cab-over-engines may have been more practical and economical, because they provided the better use of chassis space and increased payload, but aesthetics have also always been a factor in truck designs. In the late 1950s, even many European customers preferred the stylish conventional truck designs. DAF catered to these customers with the introduction of the 13, 15, and 16 models—all featuring conventional styling.

Cab-overs still more popular

Despite the introduction, DAF's cab-over models continued to dominate. The DAF conventional trucks never really took off until it introduced the N series in the early 1980s, a truck known for its ruggedness. Still, the 13, 15, and 16 series DAFs did find their place in the industry.

The 13 DA was driven by a 105hp (78.2kw) diesel engine and it had a maximum payload of 5.9 tons (6 tonnes). These trucks could tackle many different applications and were commonly used as dump trucks. These were intended for mid-range hauls. A 13 BA powered by a 135hp (100.6kw) Hercules engine provided 5.9 tons (6 tonnes) of carrying capacity.

The 16 DD truck was also powered by that same 105hp (78.2kw) engine. Commonly used as a dump truck, it boasted an increased payload of about 7.8 tons (8 tonnes), which distinguished it from the 13 DA.

The 16 BB could offer slightly more power—it was matched with a 155hp (115.5kw) Hercules gasoline engine. It could be fitted with a dump box or other bodies, and had a payload of 7.8 tons (8 tonnes).

Specifications

Country: Netherlands
Year manufactured from: 1957
Engine: Hercules
Transmission: six-speeds to bevel
Payload: 24.3 tons (up to 24 tonnes)
Applications: construction, dump, long-distance, on-highway
Special features: nicknamed the "Torpedo" because of its design.

BORGWARD

1957 Germany

BORGWARD B 544

In postwar Germany, Allied forces initially prohibited Bremen Borgward from building vehicles. The need for commercial vehicles, however, was so high that Borgward introduced a new B 3000 with a single rear drive.

Right: *Borgward concentrated on building a small range of normal-control trucks which had a choice of gasoline or diesel engine.*

Specifications

Country: Germany
Year manufactured from: 1957
Engine: six cylinder diesel
Transmission: five-speeds to bevel
Payload: 4.4 tons (4.5 tonnes)
Applications: dropside platform, dump
Special features: new diesel six cylinder engine

Carl Borgward began his working life making radiators for the Bremer Hansa Lloyd Company. He took over the company in 1921 and entered the light utility vehicle market. For the car market, Borgward produced a range of Hansa 1100 and 1400 vehicles. During World War II the most famous truck they built was the B 3000, a 2.95-ton (3-tonne) payload vehicle with a four-wheel drive. It was a reliable truck that the German army adopted on a small scale. In 1944, however, a massive Allied air attack destroyed the main factory. Nothing was left standing and Carl Borgward, as a supplier to the Nazis, was arrested after the German surrender and served 2½ years in prison.

The B 3000 was authorized for production again, but only as a 4x2. A six-cylinder 85hp (63kw) gasoline engine powered the truck, but a diesel was available too. In 1950, the 4-ton (3.9-tonne) market segment was reached with the B 4000. It adopted the same design as the postwar B 3000, with the same 280ci (4.6-liter) diesel engine.

A few years later, a new designation was applied. The B 4000 was split into two different variants: the B 533 was a 3.44-ton (3.5-tonne) model with the 303ci (4.96-liter) engine boosted to 95hp (70kw) and the B 544 was a 3.94-ton (4-tonne) model with a new 304ci (4.99-liter), 110hp (82kw) engine.

GMC
1958 USA

GMC DFR 800, DLR 800

GMC was the large truck producer in World War II, contributing more than 550,000 trucks to the cause. Following the war, it would return its focus to passenger and commercial vehicles designed for civilian use.

Specifications

Country: USA
Year manufactured from: 1958
Engine: Detroit Diesel two stroke
Transmission: 10-speeds to bevel
Payload: GCWs up to 76,000lb (34,467kg)
Applications: long-distance, on-highway, local pickup and delivery
Special features: GMC contributed more than half a million vehicles to the war effort, and the DFR and DLR 8000 models were among the first of the company's postwar commercial vehicles.

The DFR 800 was a slab-fronted cab-over-engine and it was released in conjunction with a cab-forward model—the DLR 8000. The DFR 800 had a 28in (71cm) bumper to front axle set-back, while the DLR version had 50in (127cm).

Air suspensions

Both were rated at 76,000lbs (34,467kg), gross combination weight. That included double train-trailer combinations, which were now permitted in many U.S. states. Both trucks had aluminum tilt cabs and air suspensions and were available in a 48in (122cm) bumper to back of cab (BBC) dimension. The air suspension enabled these trucks to keep the same height regardless of the load they were carrying. It also enabled the fifth wheel and trailer floor to be 3in (7.6cm) lower than in comparable configurations, which increased payload by about 70 cubic ft (2 cubic m) in a 35ft (10.6m) trailer.

By the time 1960 rolled around, GMC was an unrivaled producer of trucks. What made the design of these trucks stand out was the I-section frame rails. By 1963, GMC has reverted back to the more traditional side-members and steel leaf suspension.

Above: *DF type trucks of the 1950s era were powered by Detroit Diesel two-stroke engines which made a very distinctive sound. The heaviest-duty models had gross weight ratings of up to 76,000 lbs (34,467kg).*

MITSUBISHI

1959 Japan

MITSUBISHI FUSO T380

Mitsubishi was able to provide heavy-duty trucks for long-distance transportation with the introduction of some new highway rigs in the late 1950s. One such offering was the T380 introduced in 1959.

Above: *The Mitsubishi T380 was used to transport loads over medium to long distances in Japan. It was also the first tilt-cab to be introduced in Japan.*

Specifications

Country: Japan

Year manufactured from: 1959

Engine: Mitsubishi engine up to 165hp (123kw)

Transmission: five-speeds to bevel

Payload: up to 7.8 tons (eight tonnes)

Applications: medium- and long-distance, on-highway

Special features: Japan's first tilt-cab truck.

The T380 played a significant role in Japan's improving economy and became known as a durable and efficient truck. It was powered by a 165hp (123kw) diesel engine.

But while 7.8-ton (8-tonne) loads were considered large at that time, there was soon a need for even more powerful trucks. As a result the T390 was introduced the same year, giving fleets the ability to haul 11.3-ton (11.5-tonne) loads. It was also a tilt-cab model and at the time it had the largest carrying capacity in its class.

The T390 featured a 6x2 drive format and was driven by a powerful 220 hp DB34 turbocharged diesel engine. The T380 and T390 would pave the way for Mitsubishi to introduce its modern line of heavy-duty trucks. They also preceded the introduction of the T386 tractor unit capable of weights of 26.5 tons (27 tonnes) loaded and the T386S, which could come in at an impressive gross combination weight of 32 tons (32.6 tonnes).

New loading system

A couple of years after the introduction of these heavy-duty trucks, Mitsubishi pioneered an innovative loading system. The forward section of the truck was tilted up on hydraulic jack-legs similar to those found on cranes. Then, an already-loaded cargobed could be winched onto the back of the truck.

The T380 and T390 enjoyed much success before being replaced. The T380 met its end in 1969 when replaced by the T387 and T388 series. The T390 was replaced the following year by the T951.

VOLVO

⚒ **1959 Sweden**

VOLVO LAPLANDER

Designer Nils-Magnus "Måns" Hartelius is well known within off-road vehicle circles as the creator of a number of the cross-country vehicles that Volvo has produced. One such truck was the Volvo Laplander.

Above: *The Swedish armed forces held a competition to see who could design the best off-road vehicle. The winner was Nils-Magnus "Måns" Hartelius, the designer of the Laplander produced by Volvo.*

The Laplander came about as a result of a competition run by the Swedish armed forces. They wanted an all-purpose vehicle to replace their aging General Purpose (GP) vehicle, which was losing its efficiency.

In 1959 preproduction versions of the Laplander were presented to the Swedish military for evaluation. It was dubbed the P2304 for test purposes. The truck was a forward-control vehicle and it met a number of objectives. It decreased the length of the vehicle, increased its handling performance over rough terrain, and improved the weight distribution between the axles. It also provided the driver with excellent visibility.

After receiving input from the military and fine-tuning the vehicle, the L3314 was born. It was similar to the P2304 test vehicles, but it was now driven by a more powerful engine.

Volvo car parts involved

The L3314 Laplander was comprised largely of Volvo car parts. Components such as the gearbox and the front and rear axles were borrowed from Volvo's cars, such as the PV544 and Amazon. However, the tires were nothing like those found on passenger cars.

They needed to be extra-large to provide plenty of ground clearance. A differential lock was also added to the rear axle so the Laplander could negotiate its way out of muddy terrain.

The military version of the Laplander was open-topped with a canvas roof. However, a hardtop version was later introduced. It featured a steel cab structure with room for eight people.

Civilian version introduced

When the civilian version of the truck entered production, the hardtop version became the standard Laplander. Soon thereafter another version was released, which was better for haulage applications. It only had seating for two, but it had a small platform that could be used to transport goods. That version of the Laplander was often used for snowplowing and forest-fire fighting.

Despite its usefulness to civilians, it was the military that got the most use out of the Laplander. The vehicle proved to be effective as an antitank gun transporter for use in Norway and Sweden. The Swedish military fitted the truck with a strong rollover bar to increase crew safety.

Replaced by C3 version

The Laplander remained popular with both military and civilian users until production ceased in 1970 in favor of the more powerful C3 generation of high-mobility vehicles. The C3, however, was extremely expensive in comparison and not generally affordable to the general public. The Laplander had developed a loyal following among the civilian population and these customers continued to demand a vehicle of its type.

So, after several years of being badgered by civilian customers, Volvo resumed production of the Laplander. The Hungarian manufacturer Csepel Auto formed a partnership with Volvo to help with its production.

Specifications

Country: Sweden
Year manufactured from: 1959
Engine: four cylinder gas
Transmission: five-speed
Payload: 1 ton (1.016 tonne)
Applications: off-highway, cross-country, military
Special features: originally inspired by the military, this truck developed a strong civilian following.

Below: *The United Nations used the Laplander on humanitarian missions, because it enabled workers to traverse rough terrain and poor-quality roads.*

LEYLAND

🛠 **1960 UK**

LEYLAND OCTOPUS

In 1960 the Leyland Octopus received a number of enhancements including improved styling. The truck was commonly used to haul items aboard a flatdeck or as a tanker truck.

Above: *The Leyland Octopus was known for its contemporary styling and also its power. This advertisement plays on the fact that the truck boasted a powerful engine ideal for heavy hauling.*

Specifications

Country: UK

Year manufactured from: 1960

Engine: six cylinder

Transmission: five-speeds to hub reduction axles

Payload: not available

Applications: long-distance, on-highway, flatdeck, tanker

Special features: rubber mountings provided superior ride quality.

Above: *The Leyland Octopus was known for its lightweight cab construction, being composed of fiberglass rather than the traditional material, steel.*

The Pressed Steel Vista Vue Cab was built on a steel subframe comprised of welded box sections. Flexible rubber mountings were used to mount the cab onto the chassis, which reduced vibrations and led to a smoother ride. Ease of entry was also improved with a step height of just 18in (46cm) from the ground.

A wrap-around windshield provided excellent visibility while the rear bulkhead incorporated three windows. This enabled the driver to see what was happening behind the cab.

Stylish paneling ran along the full width of the cab and important gauges, switches, and instruments were positioned in the center of the dashboard. All of the parts and components on the Leyland Octopus were treated to resist rust and corrosion, making it a long-lasting and durable truck.

Lightweight cab

While Leyland built the Octopus with driver comfort in mind, it also made it lightweight for improved efficiency. GRP moldings replaced sheet steel pressings. The cab was identical in appearance to its steel counterpart and Leyland insisted the GRP cab was just as durable. The cab also had a lower height than other models.

GMC

🛠 **1960 USA**

THE NEW GMCS

Above: *The GMC L5000 tilt-cab model had a 180 bhp V6 gasoline engine. The heaviest L-range models had the option of a GM V-71 two-stroke diesel engine.*

In 1960 GMC completely redesigned its popular line of light-duty trucks. While light-duty in size and category, these trucks were routinely put to use hauling freight or working on North American farms.

Right: *The newly designed aluminum tilt-cab cruiser was nicknamed the "Crackerbox."*

Specifications

Country: USA
Year manufactured from: 1960
Engine: GMC V6-V12
Transmission: 10-speeds to bevel
Payload: not available
Applications: local pickup and delivery, long-distance, on-highway, fire
Special features: radically new design made these trucks stand out.

The new design was radically different than its predecessor in both appearance and functionality. The new-look GMC featured a new front and rear suspension for an improved ride, and also boasted more power. They were equipped with a standard V6 engine as of 1960.

Heavy-duty cousins

While the American public was raving about the new-look light-duty GMCs, its larger cousins were also being put to work in numerous applications. The L-model was a 72in (1.8m) steel tilt-cab truck powered by a choice of engines ranging from a GMC V6 to a V12. The truck didn't change much from year to year and was discontinued in the late 1970s. The cab-over-engine trucks were commonly customized as fire trucks.

Crackerbox

At the same time, GMC was also introducing a new aluminum tilt-cab cruiser, which was nicknamed the Crackerbox. It was introduced in 1959 and sold until 1968 when it was replaced with the Astro.

The lightweight Crackerbox boasted an extra 1,824lbs (827.2kg) of payload compared to other similar trucks and measured only 48in (1.2m) from bumper to the back of cab (BBC). This short BBC also made it ideal for maneuvering in tight places.

The Crackerbox was renamed the F-model in 1960 and was available with diesel engines or GMC's own gasoline-powered engines. The truck was still known for its light weight and high payloads, however, and a sleeper was available. It measured only 22in (56cm) deep behind the driver.

The Legend

chapter 6

Boom Time

In the 1960s, trucks overtook the railroads as the principal means of freight transportation. Both in North America and in Europe, highway networks were being expanded to accommodate heavier traffic and the quality of the roads was increasing dramatically. Britain's first highway was opened in 1959, making long-distance freight haulage by truck more practical than ever before.

Above: *The Volvo N86 and N88 may have looked like their predecessors but they were in fact completely new trucks.*

Left: *The Peterbilt 359 has a devoted following; it is a firm favorite at North American truck shows.*

It was a time of experimentation among truck designers. Volvo introduced its first tilt cab—the L475 Raske—in 1962. The tilt-cab concept was borrowed from the American trucks of the day, which had already adopted the design. It allowed easier access to the engine and other components. In the U.S., cab-over-engines were becoming increasingly popular in the late 1950s and early 1960s. Legislators were limiting the overall length of combination vehicles and in order to maximize payload, the cab-over made more economic sense. This style of truck also provided improved visibility for drivers. Other technological advances that found a home in commercial trucks during this era included synchronized gearboxes, taper leaf suspensions, fail-safe spring brakes in the U.S., and the introduction of spring brakes in Europe.

In addition to the many technological breakthroughs, manufacturers in the 1960s were still aware that esthetics were equally important and they strove to design trucks that would appeal to the driver and not just the owner. In 1964, 1.5 million commercial vehicles were sold in the U.S. alone. In the mid-1960s, eight million Americans made their living in the trucking industry in one capacity or another—whether it was building the trucks or driving them.

VOLVO

1960s Sweden

VOLVO L46/L47

The L46 and L47 were the last conventional medium-duty trucks produced by Volvo in Sweden. Both were a huge success, capable of taking on jobs many truckers would never dream of inflicting on a medium-duty vehicle.

They replaced the L36/L37 in the early 1960s and incorporated a variety of new technologies, including power steering and the revolutionary Volvo safety cabs. The buzzwords at Volvo were "active safety" and "passive safety" and both were enhanced thanks to these and other improvements.

The L36/L37 were a popular choice in the 1950s when medium-duty trucks typically traveled short distances and were rarely required to pull a trailer. As a result, a large percentage of these vehicles were produced with gasoline engines. Regardless, a number of these trucks did perform heavy-duty applications. Many were used as tipper trucks and others were equipped with cranes. As the severity of the work required of these medium-duty trucks increased, it was clear the industry was ready for a more powerful replacement, and thus the L46 and L47 were born.

More power

While the L46 and L47 may have looked like medium-duty trucks, they certainly did not lack power. The L47 was available with a turbocharged engine, even though this technology was still relatively new to the trucking industry at this time.

The L47 was included in Volvo's System 8 global truck family where it was referred to as the N84, and it was widely available throughout the world. The main difference was that the N84 had an even more powerful engine with more capacity. The extra power was welcomed because road infrastructure around the globe was improving during the 1960s, allowing for faster average speeds.

Above: *The L46 and L47 represented Volvo's last kick at the can when it came to conventional-style trucks in Europe. They came equipped with turbocharged engines, which made the trucks capable of heavier-duty tasks than they would appear at first glance.*

Specifications

Country: Sweden
Year manufactured from: 1960s
Engine: Volvo turbocharged six cylinder
Transmission: five-speeds to bevel
Payload: not available
Applications: local pickup and delivery, severe-duty, dump, crane
Special features: gasoline engines under the hood.

KENWORTH
1961 USA

KENWORTH W900/K100

Two legendary Kenworth trucks were introduced in 1961—the W900 conventional and the K100 cab-over-engine. The K100 was designed to maximize cargo within state truck-length restrictions.

Above: *Kenworth's W900 (seen here on the left) made famous the long, square hood that is still popular today among North American owner-operators.*

Specifications

Country: USA
Year manufactured from: 1961
Engine: Cummins or Caterpillar six cylinder diesels
Transmission: 10-speeds to bevel
Payload: not available
Applications: long-distance, on-highway, vocational, off-highway, oilfields
Special features: tilt hood replaces the previously common butterfly-style hood.

The Kenworth W900 featured a classic look that was still popular among North American truckers, with a long hood and distinctive cathedral grille.

It found itself being thrust into all sorts of applications from rugged off-highway work to long-distance general-freight haulage. The truck could be customized for a variety of tasks, making it a diverse vehicle, and also earned a reputation for being strong and reliable.

The W900 was also designed with driver comfort in mind, which won it fans among the over-the-road crowd. The most common engine on the original W900 was a 300hp (223.7kw) Cummins turbocharged six-cylinder NTC-300. In 1982 a modified W900B was introduced.

The W900 gained plenty of attention when one was called on to deliver a high-resolution spectrometer magnet, 36,000 times stronger than the Earth's magnetic field. The load was 140ft (42.7m) long and weighed 105.3 tons (107 tonnes), but this customized W900, with a 450hp (335.6kw) Caterpillar engine and a Spicer 24-speed transmission, was able to transport the unusual shipment all the way from Illinois to Palo Alto, in California.

K100

The K100 was an important addition to Kenworth's fleet as a result of length limitations in eastern states. Operators who insisted on running a long-nose conventional truck in those regions would be forced to sacrifice payload, whereas the shorter K100 allowed fleets to maximize their carrying capacity. Of course the aerodynamics left something to be desired and in 1984 Kenworth introduced the K100E cab-over-engine, which improved air flow.

ZIL

1961 Soviet Union

ZIL-164A

The ZIL-164 replaced the ZIS-150 on Soviet assembly lines in 1957. Four years later the truck received a number of slight changes and was renamed the ZIL-164A.

Specifications

Country: Soviet Union

Year manufactured from: 1961

Engine: 97hp (72.3kw) six cylinder

Transmission: five-speeds to bevel

Payload: not available

Applications: military, dump, cargo transportation

Special features: still commonly used for humanitarian and military operations in Third World countries.

The ZIL-164A was generally powered by a 97hp (72.3kw) six-cylinder engine capable of speeds of 46.6mph (75km/h). It weighed 9,040.5lb (4,100kg) unloaded and had a five-speed gearbox.

The original ZIL-164 and the GAZ-51A served as the predominant truck in the USSR for the first half of the 1960s. The main difference between the ZIL-164 and its ZIS-150 predecessor was the fact the ZIS-150 had a horizontal radiator grille whereas the former had a vertical one.

Beneath the surface, however, there were a number of other differences between the trucks, mainly pertaining to functionality. The ZIL-164A brought forth even more changes including a better transmission, new braking system, and telescopic shock absorbers on the front wheels. The ZIL-164A provided the basis for a wide variety of trucks performing in various applications. Many were dump trucks while others were used to haul general freight. Still others were used in the military. In fact, older versions of the truck served in the military well after being replaced on the production lines in 1964.

More ZILs introduced

At around the same time the ZIL-164A dominated Russian roadways, a number of other derivatives of the redoubtable truck were also in use, such as the ZIL-MMZ-164AN tractor-trailer, the ZIL-MMZ-585L construction dump truck, and the ZIL-MMZ-585M farm truck.

Above: *The functional ZIL family of trucks was highly popular in the Soviet Union. Here, a ZIL-130 is pictured transporting an extremely heavy load for the truck's size.*

BROCKWAY
1958 USA

BROCKWAY HUSKIE

Brockway originally introduced its Huskie model in 1958, but the line was expanded in 1961 to include more options. Brockway—owned by Mack—offered trucks that borrowed heavily from Mack features.

Above: *The front fenders were attached to the truck with two bolts that could be easily removed, providing ready access to the engine and components. This simplified maintenance made the truck popular among mechanics as well as drivers.*

Right: *The Brockway Huskie, owned by Mack, was an instant hit among American trucking enthusiasts, thanks to its abundance of chrome and high horsepower engine.*

Specifications

Country: USA
Year manufactured from: 1958
Engine: full range of Brockway, Continental gas and Cummins diesels
Transmission: not available
Payload: GVWs up to 60,000lb (27,210.8kg)
Applications: long-distance, on-highway
Special features: one of the few Class 8 trucks to be paired with gasoline engines built by Brockway with up to 200hp (149.1kw).

The new Huskies introduced in 1961 had short dimensions from bumper to back of cab (BBC) of just 90in (2.2m). Cargo trucks from this range had gross vehicle weights (GVWs) ranging from 23,000–36,000lb (10,430.8–16,326.5kg). Tandem-axle models, meanwhile, had GVWs of up to 50,000lb (22,675.7kg), and Huskie tractors offered combination weights to 60,000lb (27,210kg).

Redesigned cab

Huskie owners could choose from a variety of engines including a 478ci (8100cc), six-cylinder gasoline engine built by Brockway that offered up to 200hp (149.1kw). For those who didn't need that much power, Continental and Cummins engines could have ratings up to 160hp (119.3kw).

The new range of Huskies featured a redesigned cab, which distinguished them from the original trucks introduced three years earlier. The new cab was immediately identifiable due to its all-metal composition and its flat, one-piece windshield. In an attempt to improve access to the engine, designers attached the front fenders with two bolts, which could be easily removed by mechanics wanting to service the engine or perform maintenance.

The next year Brockway turned 50 years old and it celebrated this accomplishment by gold-plating its radiator mascot. The company then launched an ad campaign declaring "It's the year of the Golden Huskie!" It also dubbed the new Huskie the "most advanced" in the company's history.

VOLVO
1962 Sweden

VOLVO L475 RASKE TILT CAB

In North America, customers were beginning to show an increased interest in cab-over-engine trucks during the late 1950s and early 1960s, particularly because legislation was limiting the overall truck lengths.

Above: *Volvo's choice of tilt cab was influenced by American practice. It brought the benefits of improved engine access and greater comfort by insulating the driver.*

Left: *Volvo borrowed the tilt-cab concept from the U.S. and implemented it on its L475 Raske. The TIPTOP cab was the first of its kind in Europe. It was also the first Volvo to be offered only with a turbocharged engine.*

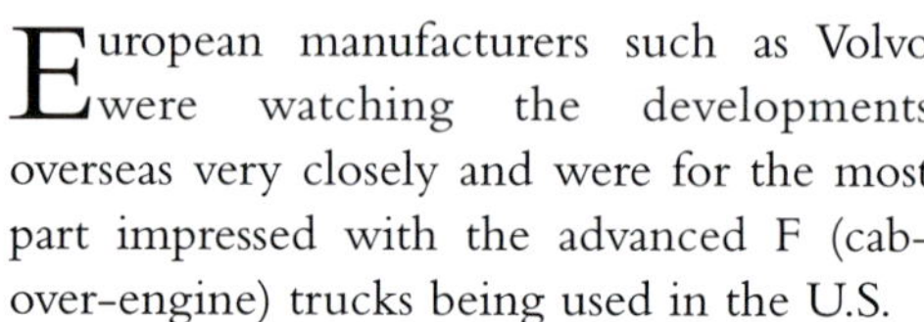

European manufacturers such as Volvo were watching the developments overseas very closely and were for the most part impressed with the advanced F (cab-over-engine) trucks being used in the U.S.

The American F trucks were extremely lightweight and efficient. But one of the most notable developments in the U.S. was the creation of the tilting cab that provided drivers and mechanics with easy access to the engine. Volvo liked the concept, so much so that it decided every new generation of Volvo heavy-duty trucks would include F trucks in all weight segments and a tilt cab would be introduced to the European marketplace.

In 1962, Volvo became the first European manufacturer to mass produce a tilt-cab truck. The truckmaker's design team in Gothenburg, Sweden, worked on the new concept under the direction of Sigvard Forssell. The Nyström cab factory in Umeå also worked closely with Volvo engineers on the project. Volvo soon decided on a suitable name for its latest innovation—the TIPTOP.

Raske debuts TIPTOP

The L475 Raske was the first Volvo truck to showcase the new TIPTOP and its introduction took place in 1962. The Raske was an ideal truck for the job because it was comprised of proven, reliable components. The new version was called the Volvo L4751 Raske TIPTOP.

The Raske was well known for its performance as well as its efficiency, so it seemed like the logical truck to experiment with. It was also the first truck to be offered only with a turbocharged engine. It featured power steering as standard, which was uncommon at this time. All in all, it was a revolutionary truck.

Customers immediately embraced the vehicle and it was reintroduced in 1965 as part of Volvo's System 8 family of trucks.

Specifications

Country: Sweden
Year manufactured from: 1962
Engine: Volvo turbocharged diesel
Transmission: five-speeds to bevel
Payload: not available
Applications: long-distance, on-highway
Special features: Volvo debuts its own tilt cab, known as the TIPTOP. Power steering was also a standard spec.

FODEN
1962 UK

FODEN TILT CAB

In 1962 Foden became the first UK-based truck manufacturer to introduce a tilt cab. The truckmaker did this with its S.24 cab, which made its debut in 1962 thanks to the advent of glass-reinforced plastic (GRP) cabs.

Above: *Foden introduced a tilt cab in 1962, so becoming the first UK manufacturer to do so. The S.24 fiberglass cab had ultramodern styling for its time.*

Specifications

Country: UK
Year manufactured from: 1962
Engine: Gardner six cylinder or Foden six cylinder two-stroke diesel
Transmission: 12-speeds to worm
Payload: not available
Applications: long-distance, on-highway
Special features: first UK manufacturer to introduce a tilt cab.

Foden was well known for its new cab designs, and was a major user of GRP in its cabs starting in 1958 with the S.21 cab.

The twin-load

Foden's rigid eight-wheelers were particularly popular trucks in the UK, but new weight limits introduced in 1964 required manufacturers to diversify. Many began building tractor-trailer units instead. Foden itself introduced the twin-load, which was a dromedary-style eight-wheeler coupled to a single-axle semi. But this project flopped, because it was an impractical configuration given the UK length limits. Very few were ever built.

Hard times ahead

Despite the failure of the twin-load, Foden was still enjoying some plenty of success in the UK during the 1960s. The 1970s were a different story, however, and the company encountered financial difficulties, going into receivership in 1979 and being acquired by Paccar Company the following year. Foden was renamed the Sandbach Engineering Company.

This historic manufacturer did make a return to truck production under Paccar and the company once again began using its original name. Foden's Sandbach factory, however, was closed and production moved to the Leyland plant, near Preston.

MACK
1962 USA

MACK F, U & MB SERIES

Mack was extremely active in the 1960s, blessing the U.S. trucking industry with a number of popular new options. A new highway tractor (F series) was to become a major, long-running success story for Mack.

Left: *Mack's F series enjoyed much success in the North American market and was widely used for heavy-haul and oversized-load applications. More than 70,000 units were sold before the F series was retired.*

Mack's F series of cab-over trucks hit the marketplace in 1962 and remained a prominent player in the industry until 1981. During this time about 70,000 units were sold, thanks largely to the forward-thinking design that engineers built into the series.

The F series had a 50in (1.3m) BBC dimension and even when equipped with a sleeper it would come in at 70in (1.8m) from bumper to back of cab. The F series boasted a modern appearance for its time and an upgraded interior made it a hit with drivers. Under the cab, a wide array of diesel engines could be specified.

Mack MB series

Around the same time as the F series was earning respect on North American highways, the MB series was becoming a force in cities. The MB series was introduced in 1963 as a replacement to the N series and before the model was retired in 1978, more than 18,500 units had been sold. The MB series was primarily a city delivery and garbage truck—the first of its kind to be paired with a Mack/Scania engine. The MB was usually equipped with a 140hp (104.3kw), six-cylinder END 475 engine, which was built by Scania-Vabis at the time.

Specifications

Country: USA
Year manufactured from: 1962
Engine: full range of Mack diesels
Transmission: 10-speeds to bevel
Payload: not available
Applications: long-distance, on-highway, regional delivery, garbage
Special features: Mack came out with a model for each specific application throughout the 1960s.

THAMES TRADER
1962 UK

THAMES TRADER

The British Ford division launched a new truck in 1958 known as the Ford Thames Trader. With a very unusual hood, this truck remained in production until 1965, when it was replaced by the D series.

Specifications

Country: UK
Year manufactured from: 1962
Engine: six cylinder gasoline or diesel engine
Transmission: five-speeds to bevel
Payload: semitrailer GCW up to 15 tons
Applications: stake body, parcel-delivery truck, dropside platform, dump, tractor unit
Special features: very distinctive semi-forward control cab

Above: *The Thames Trader was sometimes built as a tractor unit with a small wheelbase (9ft/2.74m), but the rigid trucks were more popular.*

The first British Ford truck was sold under the Fordson name. When the production facilities moved from Manchester to the banks of the Thames River, the name "Fordson Thames" was adopted to give the trucks a distinct identity. The Thames name ran from 1957 to 1965. The Thames Trader was a modern semiforward-control truck that superseded the ET series. The range catered for loads from 1½ tons (1.52 tonnes) to 7 tons (7.11 tonnes), the latter being the highest payload Ford could offer on a four-wheeler rigid truck. The model number referred to the payload. So, the Thames Trader 15 was the light 1½-ton truck, the Model 20 was the 2-tonner, the Model 30 was the 3-tonner, the Model 40 was the 4-tonner, and the bigger Model 70 was the 7-tonner. In addition, some short-wheelbase tractors were built for hauling semitrailers. Two gasoline engines, and their diesel derivatives, were offered. The four-cylinder version was a 220ci (3.6-liter) gasoline engine, born in 1953 and produced as a diesel variant a year later. This engine was designed for the trucks up to 3 tons (3.05 tonnes). Above this weight, the other engine was a gasoline or diesel six-cylinder 330ci (5.41-liter) powerplant. With a total of 121,853 semiforward trucks built (along with 19,000 normal control cab trucks), the Thames Trader was a great success. Some of the trucks were exported or manufactured in Spain by the Ebro company or produced in Turkey by the Otosan company.

DAF

1963 Netherlands

DAF A2600 SERIES

DAF knew that its future success depended on its ability to cater to the long-distance international transportation market. Its first attempt at penetrating this market with the 2000 series proved unsuccessful.

Above: *The 2600 featured an all-new cab design fixed to the chassis rather than tilting like many of its competitors of the day.*

Specifications

Country: Netherlands
Year manufactured from: 1963
Engine: DAF six cylinder diesel
Transmission: six-speeds to bevel
Payload: GCWs up to 40 tons (40 tonnes)
Applications: long-distance, on-highway, international transportation
Special features: all-new cab design with forward entry.

Above: *Although DAF's 2000 series wasn't a very successful entry in the long-distance highway market, the 2600 aimed to address its predecessor's shortcomings. It was able to do so, becoming a major player in the international truck market.*

Not that the blame for the failure should fall entirely on DAF's shoulders. Unforeseen changes to European transportation legislation and a mistaken sense of customer needs in this segment created the downfall of the 2000 series. It lasted only two years before the truckmaker pulled the plug on the experiment.

Still, DAF wasn't deterred from forging ahead with a second attempt at cracking the international transportation market and it did so with the A2600. DAF proved it learned from its first experience when it launched the A2600 in 1963 and it established itself as a major player in the northwestern European international transportation market with this truck.

The truck featured an all-new cab design, which was fixed to the chassis rather than being tiltable. DAF has been known for its cab construction ever since. The 2600 series was expanded to include the first 6x6 tipper chassis in 1965. In 1968 it introduced an 704ci (11.6-liter), 165hp (123kw) engine. This would go on to power many of DAF's trucks in the years to come.

Further expansion

The 2600 line was continuously expanded and upgraded as it earned its reputation as one of the world's greatest long-distance trucks. The line grew to include maximum-weight rigids and articulated trucks reaching gross combination weights of up to 39.4 tons (40 tonnes). The truck was so popular that DAF had to open a new cab and axle plant in Belgium to keep up with demand. In 1970 a tilt-cab option became available.

VOLVO

1965 Sweden

VOLVO N86/N88

As a successor to its popular Viking and Titan trucks, Volvo introduced the hooded N86 and N88. These were heavy-duty conventional-style trucks that looked much like the Viking and Titan at first glance.

Above: *Volvo stayed the course with conventional design with the N88. The N88 was equipped with a powerful engine enabling it to haul heavy loads over hilly terrain.*

Specifications

Country: Sweden
Year manufactured from: 1965
Engine: 410 and 585ci (6.7- and 9.6-liter) Volvo engines
Transmission: R50 and R60 synchronized eight-speeds to bevel
Payload: not available
Applications: long-distance, on-highway
Special features: conventional design was based on Volvo's experience in the North American marketplace.

Many of the developments introduced in the N86 and N88 could be traced back to Volvo's first experiences in the North American market. Volvo entered that marketplace in the late 1950s and learned a great deal about how to design premium conventional trucks as a result.

While they may have looked like their predecessors, the N86 and N88 were in fact completely new trucks. Engines of 410 and 585ci (6.7- and 9.6-liters) could be found under the hoods of these models. Previous Volvo trucks had been powered by engines of the same capacity, but the powerplants found in the N86 and N88 were capable of higher power outputs and had very efficient turbocharging. Volvo strove to achieve five primary goals: reliability, long service life, low fuel consumption, low weight, and the ability to be continuously upgraded. Furthermore, although being ecofriendly wasn't yet a huge concern in the middle of the 1960s, Volvo's new engines proved to be quieter and more environmentally sound than the high-emission engines that usually powered heavy-duty trucks during that era.

Synchronized transmission

Another development introduced in the N86 and N88 were fully synchronized gearboxes. These were dubbed the R50 and R60 and they were of the range-change type. These proved to be long-lasting transmissions, mainly because of the ease of shifting they provided. It also utilized a modern splitter, so each gear position would accommodate two speeds. It didn't take long for other manufacturers to adopt this approach.

VOLVO

1965 **Sweden**

VOLVO F86

Volvo's F86 was introduced as the successor to its L4851 Viking TIPTOP, featuring many enhancements over its predecessor. The truck went on to become one of the most successful Volvo trucks ever launched.

Specifications

Country: Sweden

Year manufactured from: 1965

Engine: six cylinder diesel

Transmission: eight-speed synchronized to bevel

Payload: not available

Applications: long-distance, on-highway

Special features: single passenger seat replaced traditional two-person bench seat.

Above: *Volvo's F86 succeeded the popular Viking TIPTOP. It became very popular in Britain and even as far away as Australia*

The F86's cab was very similar to that of the Raske TIPTOP, but it featured a larger engine tunnel to accommodate the 410ci (6.7-liter) engine. It also had a single passenger seat rather than the traditional two-person bench seat of the Raske.

Major export truck

Most of its components were borrowed from the N86 truck, which succeeded the L48 Viking. The F86's main improvements included a new engine, specially designed for high-output turbocharging. The new R50 eight-speed synchronized transmission was also introduced, simplifying the task of driving. Other new components included suspensions, brakes, and steering systems.

Major export truck

Trucking companies from around the world took notice of the improvements and began placing orders with Volvo. Great Britain, for instance, immediately took to the new offering. Before long, production of the F86 began in Scotland and it soon became the most popular truck in the country. It was also hugely successful in Australia.

The F86 was introduced onto the U.S. market in 1974. Because Volvo was still a fairly recent newcomer to the North American heavy-duty truck market, limited numbers of the F86 were sold there. Still, the truck went a long way toward establishing Volvo as a significant player in the North American trucking industry.

MACK

1965 USA

MACK R SERIES

Mack's R series of vocational trucks introduced in 1965 earned its place in trucking history over the course of nearly four decades. The R series became the world's number one mixer and dump truck.

Above: *Mack lived up to its tough Bulldog reputation with the legendary R series.*

Right: *The Mack R series is a legendary work truck capable of severe-duty applications. It became a bestseller, with more than 360,000 units sold before its retirement in 2003.*

Specifications

Country: USA
Year manufactured from: 1965
Engine: Mack Maxidyne
Transmission: 10-speeds to bevel
Payload: not available
Applications: mixer, dump, construction, off-highway
Special features: legendary dependability and longevity made this one of the best-selling Class 8 trucks of all time.

The R series itself had big shoes to fill when first introduced as a replacement for the popular B series. However, it was a proven winner, especially when paired with Mack's new Maxidyne constant-horsepower diesel engine and the Maxitorque transmission. The Maxidyne was in fact the first high-torque rise constant-horsepower diesel engine to hit the trucking industry.

The R series trucks were immediately recognizable because of their distinctive square front-end. Some of the R series' selling points were a weight-saving design, bolted construction, a tilt-hood, and a full range of components for customers to choose from.

The staggering number of R series trucks sold in North America made it the most popular Class 8 vocational vehicle in trucking history. Mack R series customers were fiercely loyal, largely because of the model's dependability and longevity. When the RD was terminated in 2003, it was estimated that more than 200,000 of the R series trucks built over the past 38 years were still in operation.

RB and DM

While production of the RD was halted in 2003 in favor of the Granite, two other models using the R series cab continued to be built. The RB (axle back) and DM (offset cab) remained in production since the Granite was not yet available in those configurations. However, the majority of R series trucks ceased production in 1989.

WESTERN STAR
1967 Canada

WHITE WESTERN STAR

Above: *The first Western Star trucks were built in Kelowna, British Columbia, where a great deal of attention and pride went into each and every truck.*

Western Star Trucks were introduced in 1967. The truck line was launched by White Motor Company in Cleveland, Ohio. However, production of Western Star trucks was to take place in Kelowna, British Columbia.

Above: *Western Star was originally intended for severe-service applications, such as logging in Canada. However, it quickly developed a loyal following and was also used in long-distance highway applications.*

This location was chosen as an ideal production area because these trucks were designed for the rugged task of logging in British Columbia and Alberta. They were also quickly adopted for other heavy-duty applications, such as mining and oilfield work. Even today the manufacturer plays an important role in these niche markets. It is also a popular premium highway tractor among owner-operators. In 1981 the range was rebadged as Western Star Trucks Inc. and became a fully Canadian company.

The company steadily improved the efficiency and engineering of its custom-built Class 8 trucks, making them a popular niche truck for vocational applications in particular.

Wraparound dash

Western Star experienced steady growth through the 1980s, expanding its product line and improving its market share. It was a particularly popular truck in Western Canada, where the company had its roots. One of the first major developments at Western Star was the move to increase the headroom in the cab to improve driver visibility. Since Western Star trucks were often operated off-road and in tight quarters, providing excellent visibility was vitally important. Western Star soon became known for its superior driver visibility and spacious cabs. Another trademark of the truck was a newly designed wraparound dashboard, which provided an individualistic look.

Western Star "down under"

Western Star entered the Australian market in 1983 with local production taking place in Brisbane. In 1991, an Australian, Terrence E. Peabody, bought the company and he successfully floated it on the Toronto Stock Exchange three years later. Australia's family of Western Star trucks have been used in various applications throughout the world ranging from the oilfields of the Middle East to military duty in Kuwait, not to mention the rugged Australian Outback. Like their North American counterparts, the Australian Western Star trucks were custom-built with durability in mind.

Supertilt hood

In 1986 Western Star introduced the Supertilt hood which further endeared it to the owner-operator market in both highway and off-highway applications in North America. It was one of the first sloped hoods to be embraced by the North American trucking industry and the same lines can be seen in today's Western Star truck lineup.

The following year Western Star introduced the Cornerstone chassis which reduced the cost, weight, and complexity of the truck. Then in 1990 Western Star introduced its 6900 model, which was designed specifically for the most demanding and rugged applications. The truck was soon adopted by the Canadian military and also became popular in British Columbia's municipal truck fleets for highway maintenance work. In addition to its range of more traditional trucks, Western Star produced cab-over-engine trucks as well, which were built in the United Kingdom by what was known as ERF.

Modern lineup

The company introduced its current lineup of 4900 series trucks in 1996 before being acquired by Freightliner in 2000. At this time, production of Western Star trucks was moved to Portland, Oregon. However, the trucks still reflect their Canadian roots and continue to be popular there. In addition, many components of the truck ranging from small plastic dipsticks to the lightweight sleeper boxes are still produced in Kelowna and shipped to Portland for assembly.

Freightliner has since expanded its dealer network in the U.S. to about 300 and has marketed the range aggressively. Today, the company's slogan is "You can't love a Western Star too much."

Specifications

Country: Canada

Year manufactured from: 1967

Engine: six cylinder American diesel

Transmission: nine or 15-speeds to bevel

Payload: not available

Applications: long-distance, on-highway, vocational, logging, mining, off-highway, oilfield

Special features: durable frame made it ideal for demanding off-road applications.

Left: *Western Star became known for its superior driver visibility and spacious cabs.*

PETERBILT
1967 USA

PETERBILT 359

In 1967 Peterbilt introduced the 359—a truck that would develop a loyal following of truck enthusiasts and is to this day a fixture at most North American truck shows.

You could argue it was the exterior that won the Pete 359 a devoted following. The truck featured a wide, long hood and wide grille that was an instant hit with truckers. It was only available with a tilt hood, unlike previous Petes, which featured a butterfly-style hood that was still the hood of choice among many drivers. The Peterbilt 359 really turned heads and was a big success.

Rapid expansion

Still, owner-operators who took a great deal of pride in the appearance of their rigs embraced the new truck and by 1969 Peterbilt had to construct a new factory in Madison, Tennessee, to keep up with the demand. The company had built 21,000 trucks during the 1960s, which was four times the number produced during the previous decade.

Many engine choices available

Peterbilt offered engines from manufacturers such as Caterpillar, Continental, Cummins, and Detroit Diesel. A range of engines, including V6s, V8s, straight-sixes, straight-eights, and V12s, were available under the typically large Peterbilt hoods through the 1960s.

The most common engine specification was the Cummins NH 220 diesel, which was typically paired with a Spicer 12-speed transmission. Typical axles during this era at Peterbilt were Timken full-floating hypoid or double reduction, and gross vehicle weights were up to 84,000lb (38,095kg). Gross combination weights were a staggering 250,000lb (113,376.6kg) at the time.

The Peterbilt 359 was incredibly popular until being overshadowed by the 379, which was introduced in 1977.

Above: *The Peterbilt 359 established itself as a favorite among North American owner-operators. Drivers found its long hood, wide chrome grille, and spacious interior appealing.*

Specifications

Country: USA
Year manufactured from: 1967
Engine: full range from Caterpillar, Continental, Cummins, and Detroit Diesel
Transmission: nine, 12 or 15-speeds to bevel
Payload: GVWs up to 84,000lb (38,095kg)
Applications: long-distance, on-highway
Special features: first Peterbilt to feature a tilt hood.

DAIMLER-BENZ

1967 Germany

THE DÜSSELDORFER

The Düsseldorfer was introduced in 1967 as the replacement for Daimler-Benz's L 319. At the time, the vanlike vehicle was seen as huge step forward in truck engineering.

Specifications

Country: Germany

Year manufactured from: 1967

Engine: 244ci (2-liter) Mercedes-Benz up to 55hp (41kw)

Transmission: five-speeds to bevel

Payload: GVWs up to 4.5 tons (4.6 tonnes)

Applications: local pickup and delivery and municipal duties

Special features: had an engine borrowed from Mercedes-Benz passenger cars.

Right: *Half van, half flatbed truck, the Düsseldorfer was widely recognized as a huge step forward in truck engineering.*

In addition to being dubbed the Düsseldorfer, the vehicle was also known as the L 406 D, its official model designation. The Düsseldorfer had permissible gross weights ranging from 3.4 to 4.5 tons (3.5 to 4.6 tonnes) with a 244ci (2-liter) engine capable of 55hp (41kw). The engine was borrowed from the Mercedes-Benz 200 D passenger car.

The series was expanded by Daimler-Benz and by 1977 it included models up to the L 613 D with a gross weight of 6.3 tons (6.5 tonnes). This larger version was driven by a 130hp (96.9kw) six-cylinder in-line engine, which had its origins with the LP series. Some alternative models could be fitted with gasoline engines but they were not standard.

One of the features that made the Düsseldorfer immediately recognizable was its slight front-end bulge, which could be compared to a rather subtle hood. This was to accommodate the engine, which had to be housed partially in the cab to save space.

Cargo van line expanded

After taking over of Hanomag-Henschel, the heavy-duty cargo van line underwent expansion. Front-wheel drive Hanomag vans with gross weights of up to 3.4 tons (3.5 tonnes) complemented the Düsseldorfer. These new vehicles were modified to reflect more of the Mercedes-Benz characteristics and were badged the L 206 D and the L 306 D.

Meanwhile Henschel trucks (also involved in the takeover) were further broadening the company's truck line. Both the vans and trucks were equipped with Mercedes-Benz components.

DAF

1969 Holland

DAF 1600-2200

In 1969 DAF introduced its 1600, 1800, 2000, and 2200 series of flat-fronted, tilt-cab trucks. In 1972 the slightly narrower 1200 and 1400 series appeared, and the 2800 series introduced in 1973 rounded out the tilt-cab family.

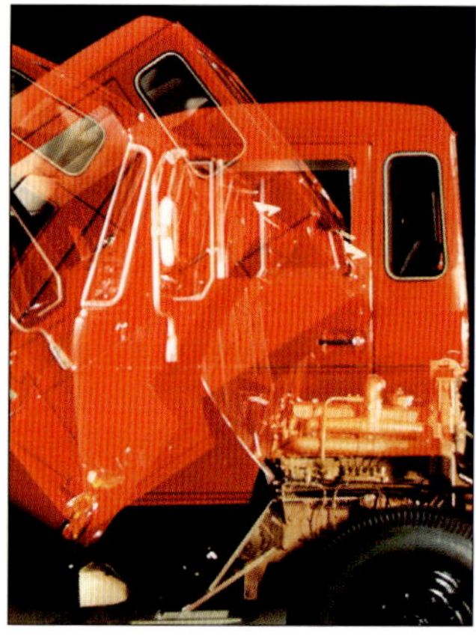

Above: *The tilt-cab was flat-fronted and offered incredible access to the engine and other components.*

Left: *A turbocharged engine could be found under the cab of most DAF 1600-2200 models. This family of trucks was available with or without a sleeper cab.*

All of these trucks were powered by 704ci (11.6-liter) turbocharged engines with the exception of the lightest version 230hp (171.5kw) engine. The heaviest of the trucks featured intercooling. Two cabs were available on these trucks—one was a long cab with a sleeper compartment and two beds and the other was a daycab with no sleeper. DAF prided itself on the comfort of its cabs.

The 1600 series also included a special version, the V1600, which had four-wheel drive and was intended for off-highway construction work. It had a payload of 5.9 tons (6 tonnes) and was powered by a 120hp (89.4kw) engine.

The DAF NAT 2505

In 1973 DAF's NAT 2505 6x4 made its debut with a chassis and cab borrowed from International Harvester. International Harvester had purchased one-third of the company's shares the previous year and the two companies began sharing parts and components.

In 1974 DAF introduced its first four-axle truck, the FAD 2200 8x4. It was destined for the UK truck market. The following year DAF trucks featured the Club of Four cab. The cab resulted from a development program involving DAF, Magirus Deutz, Saviem, and Volvo. This program was also known as the European Truck Design and it was based on the lightweight models F700 and F900.

DAF encountered financial troubles in 1975 after rapid expansion and investments in new facilities. Dutch State Mines stepped in and bought 25 percent of DAF's shares.

Specifications

Country: Holland

Year manufactured from: 1969

Engine: DAF turbocharged six cylinder

Transmission: six or 12-speeds to bevel

Payload: not available

Applications: long-distance, on-highway, construction, off-highway

Special features: comfortable cab, some of which had two bunks.

FORD

1970 USA

FORD L AND W SERIES

By 1970 Ford had introduced the new L (or Louisville) range of trucks to replace the normal-control T range. At the same time, Ford brought a new heavy-duty tilt-cab model known as the W series to the market.

Right: *Ford's highly successful Louisville or L-line trucks announced in 1970 carried on in production until the 1990s. They sold well, not only in North America, but in many overseas markets.*

Specifications

Country: USA
Year manufactured from: 1970
Engine: Cummins, Detroit Diesel, Caterpillar up to 270hp (201.3kw)
Transmission: not available
Payload: not available
Applications: long-distance, on-highway, local pickup and delivery
Special features: these trucks represented Ford's last efforts in the Class 8 market

These trucks were built at what was the world's largest truck factory at the time capable of producing 336 trucks a day.

Louisville range embraced

The W series trucks featured a 52in (1.32m) BBC and had a square look to the cab. There was also an 82in (2.08m) BBC sleeper-equipped W series truck available. Four different Cummins engines could be matched with these trucks offering a total of 13 different power ratings. Detroit Diesel and Caterpillar engines were also options and most delivered between 250 and 270hp (186.4 and 201.3kw). The L range was extremely successful and was in production for more than 20 years. The L-9000 and LT-9000 in particular were embraced by the trucking industry and many considered these trucks to be the manufacturer's first real successful foray into the conventional diesel haul market.

The W series, on the other hand, received an overhaul in 1975 at which time it was given a new grille that looked much like the one found on the more popular L series. The W series was replaced in the late 1970s by the all-aluminum CL-9000.

In 1997 Ford sold its heavy-duty truck operations to Freightliner.

MACK
18
TRUCKING
CC-12944
80,000

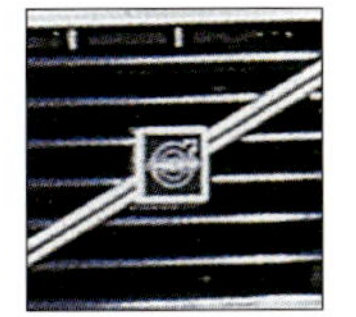

chapter 7

Economy Drive

The fuel crisis of the 1970s had dire consequences for the global trucking industry. The U.S. and its Western European allies faced a shortage of fuel as a result of a political dispute with some of the Arab members of OPEC. The major oil producers refused to sell petroleum to countries that supported Israel in its conflict with Egypt. At around the same time, OPEC quadrupled the price of oil.

Above: *The Freightliner WFT-8164 was designed so that many of its parts were interchangeable with its cab-over engine sister the WFT-7242.*

Left: *The Super-Liner could be found tackling any type of application that demanded high horsepower, including logging.*

It was the first fuel shortage suffered by the U.S. since World War II. The limited amount of fuel sent prices skyrocketing and many countries, including the U.S., were forced to introduce rationing. In a further desperate attempt to curb fuel consumption, the U.S. imposed a national speed limit of 55mph (88.5km/h).

The only way trucking companies could remain competitive was to find ways of reducing their fuel consumption. As a result, manufacturers began producing more fuel-efficient trucks. Aerodynamics were now more important than ever before and engines that delivered better fuel economy were favored. Lightweight chassis and components were also popular during the 1970s—trucking companies were taking advantage of anything that could reduce their fuel consumption. Roof air deflectors were also introduced in the 1970s to help improve efficiency. But while truck manufacturers did everything possible to make their trucks more fuel efficient, sales still plummeted.

In other developments from this era, the trucking industry in the U.S. was deregulated in the late 1970s, which presented a whole new set of challenges to the industry. Shipping rates dropped drastically as a result of deregulation and the competition between fleets was fierce.

VOLVO

1970 Sweden

VOLVO F89 AND G89

As the 1970s began, there was a need for more powerful trucks within Europe. Highways were becoming busier and it was no longer acceptable to have heavy-duty trucks traveling at slow speeds.

Germany was the first country to introduce a minimum horsepower per tonne train weight in the late 1960s. Considering Germany was a significant market for truck manufacturers, it wasn't long before European truckmakers began developing more powerful engines.

Manufacturers had to make the decision between either constructing very large and heavy, normally aspirated engines or smaller, more efficient, turbocharged ones. The normally aspirated engines would require as many as 10 cylinders and up to 1,092.6ci (18 liters), while turbocharged engines required only 728.4–849.8ci (12–14 liters). Naturally, different manufacturers chose different options, but Volvo decided to go for turbocharging.

The "Supertruck" is born

In order to meet the German regulations, Volvo developed an in-line six-cylinder turbocharged 728.4ci (12-liter) engine (the VolvoTD120A) and paired it with two new trucks—the F89 and G89. The main difference between the two was the fact the G89 had a set-forward front axle. The remainder of the truck was comprised of many existing Volvo components including the SR61 synchronized 16-speed gearbox and the NR2 rear axle with hub reduction. Soon, people were referring to the F89 and G89 as Volvo's "Supertruck."

It wasn't long before the F89 and G89 were common on European highways and in the Australian Outback where it was a popular timber transportation truck.

Above: *The Volvo F89 needed a powerful engine so its drivers could maintain reasonably high average speeds while loaded.*

Specifications

Country: Sweden

Year manufactured from: 1970

Engine: Volvo TD120A 728.4ci (12 liter) diesel

Transmission: SR61 synchronized 16 speeds to bevel

Payload: up to 52 tons (52 tonnes)

Applications: long-distance, on-highway

Special features: designed with horsepower in mind to maintain high average speeds while loaded.

VOLVO

1970 Sweden

VOLVO GLOBETROTTER

Volvo's Globetrotter cab was originally built as a "hotel on wheels" for drivers from countries behind the Iron Curtain. Volvo decided these drivers could be best served with a luxurious, spacious cab.

Above: *The Globetrotter cab raised the bar in terms of providing drivers with a "hotel on wheels." It made the truck especially popular among long-distance truckers.*

Specifications

Country: Sweden
Year manufactured from: 1970
Engine: six cylinder diesel
Transmission: 16-speeds to bevel
Payload: not available
Applications: long-distance, on-highway
Special features: designed to serve as a hotel on wheels for long-distance haulers.

Volvo claimed that the Globetrotter cab would be one of its greatest international successes. The most important feature of the Globetrotter cab was its huge amount of interior space. Volvo were also among the first to offer amenities such as a refrigerator, water tank, sink, and kitchen. Customers could specify which of these options they wanted and were then charged only for the amenities they selected.

The spacious interior was standard in all Globetrotter cabs, however. It didn't take long for trucking companies to realize there were more benefits to the large Globetrotter cab than simply extra living space for truckers from behind the Iron Curtain.

Increased efficiency

Volvo had introduced its popular F10/F12 lineup in 1977 and operators of those trucks were beginning to realize there were many benefits inherent in spacious, high-class truck cabs. Drivers who had a comfortable cab were proving to be more efficient, in addition to enjoying improved health and overall well-being while behind the wheel. This prompted Volvo to make the Globetrotter cab available on a number of its trucks for customers from all regions.

Not only did the Globetrotter enhance the working conditions for individual drivers, but it also allowed the extra space required by drivers working in teams.

SAVIEM
1971 France

SAVIEM SM 300T

Saviem was formed in 1955 by Renault trucks, Somua, Floirat, and Latil, and created a range that was both wide and popular. At the beginning of the seventies, the SM 300 was Saviem's most powerful truck.

Above: *The tilt cab with sleeper compartment was studied by Saviem and MAN. It was one of the most comfortable driver cabs in Europe.*

Above: *The SM 300T was designed for heavy haulage. The rear tandem axle was a very solid structure.*

Specifications

Country: France
Year manufactured from: 1971
Engine: MAN six cylinder diesel
Transmission: 10-speed
Payload: 24.6 tons (25 tonnes) GCW as tractor
Applications: heavy hauling
Special features: Saviem-MAN shared tilt cab, MAN engine

At first, the Societe Anonyme des Vehicules Industriels et Equipements Mecaniques was only a marketing organization. The trucks kept their original names and components under a labeled nameplate like Saviem Renault or Saviem Somua. In 1960, production was rationalized. The first real Saviem trucks were the light and popular S series. They were powered by Renault gasoline or Perkins diesel engines, because no French diesel engines were available. Then in 1962 the heavy-duty JL line was introduced. It featured a new cab and the range extended from 6.8 to 12.7 tons (7 to 13 tonnes) as rigid trucks and up to 34.4 tons (35 tonnes) for tractor units.

MAN agreement

During the sixties, Saviem forged an agreement with MAN. The German manufacturer provided its direct injection diesel engines; these were later built in France under license by Saviem. In 1970, Saviem joined the "Club of Four" along with Volvo, Daf, and Magirus-Deutz, in order to develop a modern tilt cab. The French group, however, built its own tilt cab for a new heavy-duty range, the SM series. This tilt cab was certainly the most advanced cabin in Europe and it became the standard cab on MAN as well as Saviem chassis. In 1974, Berliet joined Saviem and finally merged under a single name, Renault.

VOLVO
1970s Sweden

VOLVO C3

Volvo has earned its reputation as a leading producer of cross-country vehicles since 1928. One of Volvo's most famous cross-country models was the C3, introduced in the 1970s.

Specifications

Country: Sweden
Year manufactured from: early 1970s
Engine: six cylinder Volvo engine
Transmission: not available
Payload: GVWs up to 5.4 tons (5.5 tonnes)
Applications: cross-country
Special features: won the Paris-Dakar rally in 1983.

Above: *The C3 was designed for off-road work, but it also proved to be a winner in the desert during the 1983 Paris-Dakar rally. It won its class, earning the truck more recognition from both civilians and military personnel.*

Volvo's C3 was specially designed for cross-country applications and was widely viewed as the most efficient cross-country vehicle ever to be developed and mass produced in its class.

Race winner

The C3 gained even more fame when it took part in the famous Paris-Dakar rally in 1983. This was considered the most demanding vehicle rally in the world, and two 3.4-ton (3.5-tonne) C303s took part in the race—one of them ultimately winning the light truck category after 6,214.8 miles (10,000km) of cross-country and desert driving under challenging conditions.

When it wasn't tearing through the desert, Volvo's C3 series of trucks could often be found working as both civilian and military vehicles in applications including power production, fire fighting, and road construction. The military also used the truck as an ambulance and as a mobile base for intelligence staff.

The C3 family of trucks was available with gross vehicle-weight ratings of between 3.4 and 5.4 tons (3.5 and 5.5 tonnes). The trucks had either two or three axles and had all-wheel drive. Most of the C3s had a powerful straight-in-line six-cylinder engine with special front and rear axles. The wheels were placed lower than the center of the axles, providing plenty of ground clearance which contributed to its off-road performance.

DODGE

1972 USA

DODGE BIGHORN

Above: *The styling on American Dodge heavy trucks of the 1970s was bold and meant to convey a sense of toughness.*

Dodge began building its first Bighorn heavy-duty Class 8 trucks in about 1970, but official production didn't kick off until 1972. In 1971 about 30 Bighorns were built for a car transportation company.

Specifications

Country: USA

Year manufactured from: 1972

Engine: Cummins six cylinder diesel

Transmission: 15-speeds to bevel

Payload: GVWs up to nine tons (9 tonnes)

Applications: long-distance on-highway

Special features: this proved to be Dodge's last conventional-style heavy-duty truck.

Above: *Fewer than 300 Dodge Bighorn trucks were built before Dodge abandoned the heavy-duty truck market. They are highly sought after by collectors and truck restorers.*

The Bighorn was an 8.5-ton (9-tonne), 6x4 short cab truck and was available with three axles and a diesel engine. Between 1973 and 1975 261 Bighorn Dodge Class 8 trucks were built before the manufacturer abandoned the commercial truck industry in favor of the Ram pickup truck and other small vehicles, because it felt that the heavy-duty truck market was simply no longer profitable. Of those 261 trucks, 190 were sold in the U.S. and 70 in Canada. A lone Dodge Bighorn went to work in Mexico.

Despite their modest numbers, the Bighorn was certainly able to hold its own in the North American trucking industry. It is estimated that only 50 of these unusual trucks are still in circulation today, so they are highly sought after by collectors and restorers.

The most famous Bighorn

One 1973 Dodge Bighorn has been modified to compete in tractor pulls. Its owner, Larry Carey, enters the truck in numerous competitions in the U.S. The truck is one of the last heavy-duty Dodge trucks ever built and it is now powered by a Cummins engine capable of a massive 1,150hp (857.5kw).

MITSUBISHI
1972 Japan

MITSUBISHI FUSO T813ARA

Mitsubishi Fuso would ultimately become famous for its medium-duty Canter lineup and other less-than-large offerings. However, it did provide the industry with some heavy hitters as well, such as the Fuso T813ARA.

Right: *The Mitsubishi Fuso T810R was part of a large range of trucks offered by the Tokyo-based company and was powered by their own make of 182hp (135.7kw) six-cylinder diesel.*

Specifications

Country: Japan
Year manufactured from: 1972
Engine: Mitsubishi V10 DC6 diesel up to 375hp (279.6kw)
Transmission: five or 10-speeds to bevel
Payload: not available
Applications: long-distance, on-highway, heavy loads
Special features: MCA engines featured antipollution technology.

The T813ARA was powered by a V10 DC6 diesel engine with a power output of 375hp (279.6kw). It was the first Japanese cab-over-engine truck to integrate a hydraulically operated tilt cab. This innovation provided for simplified daily maintenance and engine servicing. It was a mechanic's dream.

The year 1972 was a significant one for Mitsubishi Fuso. In addition to launching the T813ARA, the company also demonstrated that it was truly ahead of its time by introducing the Mitsubishi Clean Air (MCA) engine series. This series included an antipollution device to reduce harmful emissions.

Previous heavy-duty trucks

While the Mitsubishi Fuso T813ARA was the first true heavy-duty highway truck offered by the company, it had previously released some other heavy-duty vehicles. The T910 for instance, had been produced five years earlier. It was a 10.8-ton (11-tonne) truck featuring dual axles at both the front and rear of the truck. It was fitted with the powerful V6 200hp (149.1kw) DC engine.

FREIGHTLINER
1973 USA

FREIGHTLINER WFT-8164 CONVENTIONAL

White Freightliner introduced its first conventional truck in 1973. It was a challenging time for the company. Fuel prices continued to skyrocket due to a shortage and truck sales were down considerably as a result.

The WFT-8164 conventional was designed so that many of its parts (about 80 percent) were interchangeable with its cab-over-engine sister, the WFT-7242, which had been introduced in 1960 and featured a 72in (1.8m) sleeper cab. The truck met the restrictive length laws in place throughout most of the U.S. at that time.

The new offering was among the lightest conventional trucks on the market, making it a popular choice for owner-operators and fleets that preferred that style of truck. The conventional truck tractor chassis weighed in at 10,315lbs (4,678kg) while a 6x4 version weighed a mere 12,465lbs (5,653kg). Cummins engines could usually be found under the hoods of these trucks.

Cab-overs not abandoned

While Freightliner had jumped head first into the conventional truck market, it wasn't about to discontinue its cab-over-engine trucks. In fact, the company introduced a new COE—the Powerliner. The main difference between the Powerliner and the company's other COE offerings was the massive new grille and slightly wider windshield. The trucks often had a huge V12 Detroit Diesel engine under the driver, although smaller engines from Cummins, Caterpillar, and Detroit Diesel were also available. The Powerliner never really took off and it was discontinued shortly after its introduction, making it a rare sight on North American highways.

Above: *With fuel prices skyrocketing, Freightliner introduced its very first conventional truck.*

Specifications

Country: USA
Year manufactured from: 1973
Engine: Cummins, Detroit Diesel, or Caterpillar six cylinder diesel
Transmission: 15-speeds to bevel
Payload: not available
Applications: long-distance on-highway
Special features: used parts common with cab-over-engine sister truck.

DAIMLER-BENZ

1973 Germany

DAIMLER-BENZ NG

Daimler-Benz launched its new generation (NG) trucks in 1973. The first NG vehicles to be introduced were aimed at the construction market, but the entire line was available within a year.

Above: *The spacious NG cab was popular among drivers, especially those who transported freight long distances and had to sleep in the bunk.*

Right: *The "New Generation" Mercedes-Benz trucks featured a completely new cab designed to look nonaggressive and to provide a high degree of driver comfort and safety.*

Specifications

Country: Germany

Year manufactured from: 1973

Engine: Mercedes-Benz up to 340hp (253.5kw)

Transmission: six-speeds

Payload: GVWs up to 22 tons (22 tonnes)

Applications: construction, long-distance, on-highway

Special features: eventually became one of the first heavy-duty commercial vehicles to offer ABS.

The NG line included two- and three-axle trucks ranging from 15.7 to 21.6 tons (16 to 22 tonnes) gross vehicle weight (GVW). A long cab and sleeper were also available for long-distance transportation.

The NG cabs were redesigned and equipped with an angled windshield and low-drop side windows for improved visibility. Engines from the LP line were initially used to power the NG trucks. A 192hp (143.1kw) V6 engine was also an option. A new feature on the NG was its planetary hub reduction axles, which transmitted high engine torque ratings allowing for the use of transmissions with a wide ratio range.

NG lineup expands

Daimler-Benz expanded its NG line to include medium-duty trucks in 1975. The NG underwent continuous development and new technologies were incorporated into the design. One example of this was the torque converter lockup clutch for semitrailer tractors. An antilock braking system (ABS) was also provided on the NG beginning in 1981, which was a huge step forward in the safety of commercial vehicles.

MITSUBISHI

1973 **Japan**

MITSUBISHI FUSO CANTER

Mitsubishi Fuso's third-generation Canter was introduced in 1973 and it boasted some substantial changes to many key components. The cab's front panel also received major modifications during the redesign.

Specifications

Country: Japan
Year manufactured from: 1973
Engine: Mitsubishi up to 100hp (74.5kw)
Transmission: five-speeds to bevel
Payload: not available
Applications: pickup and delivery
Special features: MCA system that helped reduce emissions was introduced.

The Canter's old gasoline engines were replaced with more powerful versions now offering 100hp (74.5kw). An MCA system was also incorporated to help reduce exhaust emissions—a forward-thinking move considering emission-reduction wasn't a big issue in the early 1970s.

The Canter lineup was expanded at this time to include a 2.9-ton (3-tonne) diesel truck. The Canter lineup was already known to include a wide variety of models and one more addition rounded out the lineup nicely.

The new third-generation Canter was also more flexible when it came to customization of the truck to fit a variety of applications. A number of different types of cargo loads and usage conditions could be met by customizing a model with either a low bed, high bed, or standard bed. A long-body option was also available.

Fourth generation introduced

The third-generation Canter was hugely successful and the truck saw its market share grow by more than 20 percent by the late 1970s. The introduction of the fourth-generation Canter—the FE1 and FE2—in October 1978 helped further improve the Canter's market share. This Canter was painted a color called "Arizona Cream" which was designed to resemble the color of the sun against an endless blue sky.

Above: *The Mitsubishi Canter was designed for light loads of around 2 tons (2.03 tonnes) and was powered by a choice of gasoline or diesel engine. This example has a crane fitted and is shown loading a larger truck with bamboo.*

MITSUBISHI

1973 Japan

MITSUBISHI FUSO F SERIES

Mitsubishi released its heavy-duty lineup, the F series, in 1973. It consisted of four major models including a 7.8-ton (8-tonne) class FP with a 4x2 format and a 9.8-ton (10-tonne) FT with dual front axles.

Above: *The Mitsubish Fuso FU trucks had a turbocharged engine for heavy-duty haulage. The 9.8-ton (10-tonne) FU was configured as a 6x2 drive format.*

Specifications

Country: Japan

Year manufactured from: 1973

Engine: Mitsubishi 305hp (227kw), 8DC8, Mitsubishi 265hp (197.6kw), 8DC4

Transmission: not available

Payload: GVWs of up to 9.8 tons (10 tonnes)

Applications: delivery and construction services

Special features: continuous improvements to the engine resulted in quieter ride, longer engine life.

Two years after the introduction of the initial F series lineup, a long cargo body FT212 and FU212 were introduced. These vehicles provided increased carrying capacity. In 1976 a 7.8-ton (8-tonne) FP117 was also added to the lineup. This model featured a new six-cylinder, 215hp (160.3kw) 6D20 direct-injection diesel engine, which boasted special independent cylinder heads that were designed to prolong engine life while offering better performance. A smooth, quiet operation was touted by the manufacturer. In 1977 the F series continued to grow with the addition of the four-axle FS model.

Continued expansion

Mitsubishi continued to build on their successful lineup with the addition of turbo engines in the FT and FU models in 1979. The 6D22 turbocharged direct-injection diesel engine had a maximum power output of 270hp (201.3kw), elevating the F series of trucks to the next level in terms of power and performance.

Improvements to the look of the F series trucks were implemented in 1979 with the manufacturer cognizant of changes that would be demanded by customers in the 1980s. The FP, FT, FU, FV, and FS all received a facelift on the brink of the new decade in an attempt to modernize the lineup.

Minor changes were also introduced to the FK medium-sized trucks and this line was rereleased as the new FK series. The FK and FM were the medium-duty offerings in this lineup.

MACK

1975 USA

MACK CRUISE-LINER

In the late 1970s and early 1980s, cab-over-engine trucks were still a common sight on North American highways. The W Cruise-Liner series was Mack's premium cab-over offering between 1975 and 1983.

Specifications

Country: USA

Year manufactured from: 1975

Engine: six cylinder turbocharged diesel

Transmission: 10-speeds to bevel

Payload: not available

Applications: long-distance, on-highway

Special features: semi-wraparound dashboard provides improved access to gauges.

Above: *Although Mack's Cruise-Liner cab-over had a massive, intimidating stance, it was surprisingly lightweight. The interior was enormous by the day's standards, making it a popular choice for long-haul drivers.*

Like the Super-Liner, the Cruise-Liner series was designed to offer customers a lightweight truck with huge power capability. Many of the components were lightweight, including the front springs (which also had shock absorbers attached for a smoother ride).

The trucks also featured a lightweight aluminum cab with a massive interior, which provided a great deal of comfort for the driver. A semi-wraparound dashboard provided drivers with convenient access to gauges and instruments, and the drivers' seats were air suspended and featured nylon cloth inserts for increased comfort.

Increased cooling capacity

The cab had a fiberglass-reinforced plastic roof for improved safety in the event of a rollover. It was available in configurations ranging from a nonsleeper bumper to back of cab (BBC) of 54in (1.37m) to a 90in (2.28m) BBC sleeper with a 36in (90cm) wide bunk.

The exterior of the Cruise-Liner trucks had a large radiator capable of providing enough cooling capacity for the large, powerful engines often found in these vehicles. The Cruise-Liner was adaptable to a wide range of applications and was largely customizable. Mack, Detroit Diesel, and Cummins engines were available.

KENWORTH

1976 USA

KENWORTH VERY IMPORTANT TRUCKER

Above: *The limited edition VIT series Kenworths were individually marked and each bore the name of a particular state.*

In 1976 Kenworth decided it wanted to celebrate the United States' Bicentennial year. It introduced the limited edition Very Important Trucker (VIT) series equipped with a luxurious Aerodyne sleeper.

Right: *The Kenworth Very Important Trucker series featured a luxurious Aerodyne sleeper with plenty of headroom for drivers to stand up in.*

Specifications

Country: USA
Year manufactured from: 1976
Engine: full range
Transmission: 15-speeds to bevel
Payload: not available
Applications: long-distance on-highway, COE and conventional-styles
Special features: interior comforts including 7ft (2.1m) of headroom

Both the W900 conventional and K100 cab-over trucks could be paired with the VIT interior, which offered an unprecedented level of luxury to drivers. The trucks offered standing room with 7ft (2.1m) of headroom and two luxurious and comfortable double beds. The clothes closets it had were a convenience normally associated with hotel rooms. The interior was largely leather with a diamond button tuck, which made it far more classy-looking than the typical cab interior. Refrigerators and hot plates were also included for the driver's convenience, giving truckers all the comforts of home while traveling out on the road.

To celebrate the Bicentennial, each truck bore the name of a different state. This differentiated them from the others and made the trucks true collector's items. They are still cherished by many of their owners who have kept them as show trucks once their working years were behind them.

The VIT trucks are collector's items today, but the VIT Aerodyne interior can still be customized on today's W900Ls.

VOLVO

1977 **Sweden**

VOLVO F10 AND F12

Volvo made history when it introduced its F10 and F12 trucks in 1977. The Swedish manufacturer revolutionized the trucking industry with these two models, which set a new standard for ergonomics and safety.

Above: *The F12 was capable of extremely demanding duties, such as hauling logs through the hilly Scandinavian terrain.*

To help design one of the safest commercial trucks on the road, Volvo established a special group of people who would investigate all truck accidents in Sweden and determine what went wrong. The ultimate goal was to design safer trucks and learn from the mistakes of the past. At the same time, special attention was being given to ergonomics. A more conveniently designed and comfortable truck is ultimately a safer truck. The solutions that were reached by designers and the safety team were not only implemented in the F10 and F12, but in every other Volvo truck developed since then.

Volvo opted to use chassis and driveline components that had been in use since 1973 on the N line of trucks to ensure some commonality between the models. These components had already proven themselves in a wide range of applications and the philosophy of "If it ain't broke, don't fix it" was followed when it came to these parts.

Instead, Volvo engineers directed their attention to improving safety through other modifications to the vehicle. One new development was fitting the cab to the chassis with a spiral suspension to improve driver comfort and ride. This simple change was effective in reducing the back pain suffered by many long-haul truckers.

Adjustable steering wheel

The steering wheel was made adjustable so the driver could set it at his or her preferred location and angle. In addition, the windshield size was increased so drivers had better visibility, reducing the risk of an accident. More enhancements for convenience were also introduced into the F10 and F12 models, including a luggage compartment, which was accessible from the exterior, but locked from the inside to prevent theft. This was a nice touch for long-distance truckers—the main target market for these trucks.

Injury prevention a priority

Despite its best efforts, Volvo realized accidents could not be completely eliminated. The company worked hard to make the interior of the truck as safe as possible so drivers had a better chance of escaping injury in the event of an accident.

Padding was added to the inside of the cab and virtually every sharp edge was eliminated. And, naturally, the cab was tested to the stringent Swedish cab test standards, which were the most demanding in the world. Driver health was also a motivation for the new integrated air-conditioning system. In Europe, previous air-conditioning systems consisted of air being blown through a single vent, which wasn't healthy. Volvo's new integrated system alleviated this problem and so was a popular feature of the F10 and F12.

Slightly heavier

Customers were not deterred by the fact that the cab weighed more than some of the other less safety-oriented cabs on the market. For most, the trade-off in weight was worthwhile given the safety enhancements. The F10 and F12 were developed before advanced computer-aided design techniques were available. The F10 and F12 cab featured a flat roof, but the Globetrotter cab was made available in 1979 for operators craving some extra space.

Volvo continuously enhanced the F10 and F12 over the next few years and by the mid-1990s more than 200,000 of these trucks had been sold.

Inset: *The F10 and F12 cabs weighed slightly more than other comparable cabs, but customers didn't mind because safety was enhanced so drastically.*

Specifications

Country: Sweden
Year manufactured from: 1977
Engine: 586- and 732ci (9.6- and 12-liter) Volvo
Transmission: not available
Payload: not available
Applications: long-distance, on-highway
Special features: one of the safest cabs ever produced at the time.

Left: *Volvo's F10 and F12 trucks set a new standard for safety and ergonomics. The cab was the company's safest yet, thanks in part to the formation of a Volvo Safety Team that investigated accidents and took steps to address any problem areas.*

PETERBILT

1977 USA

PETERBILT 379

The Peterbilt 379 is one of the most popular trucks among image-conscious owner-operators. Successor to the equally popular Peterbilt 359, it shared many of its classic styling characteristics.

Specifications

Country: USA

Year manufactured from: 1977

Engine: full range of Caterpillar, Cummins, and Detroit Diesel

Transmission: full range available

Payload: not available

Applications: long-distance, on-highway

Special features: blunt, long-nosed look made it an instant favorite among owner-operators.

Above: *The Peterbilt 379 was a true luxury truck that developed a strong following among North American owner-operators. Like its predecessor the 359, this truck featured a long, flat hood and a large chrome grille.*

The Peterbilt 379 appealed to truckers who wanted to enjoy a sense of luxury while behind the wheel. The truck attracted attention on highways across North America and owner-operators who took pride in their equipment dreamt of owing one. Caterpillar and Cummins engines in a wide range of power ratings were the common engine specification. Detroit Diesel powerplants were also initially available, but weren't often fitted in the Model 379. The most common application for the 379 was linehaul.

Gear fast, run slow

With the fuel crisis affecting every fleet in the 1970s, many companies were looking at aerodynamic trucks to cut down on their fuel costs. The original Pete 379 with its big, square front end was not marketed as such a vehicle.

However, Peterbilt sought other means of saving fuel and that included the motto "gear fast, run slow." By shifting at the optimum points and running the engine in the right rpm band, drivers were able to compensate for having a traditional-style truck.

Also, Peterbilt would custom-build trucks such as the 379 with the operator's specific application in mind. Matching the right components to the job was another way of reducing fuel costs.

MACK
1978 USA

MACK RW SUPER-LINER

Mack's RW Super-Liner series brought a whole new bold conventional style to the heavy-hauling sector of the trucking industry in 1978. But despite its rugged reputation, the truck also set a new standard for driver luxury.

Above: *The Super-Liner had a powerful engine under the hood, which necessitated the larger than normal front grille for cooling.*

Specifications

Country: USA
Year manufactured from: 1978
Engine: Mack, Cat, Cummins, Detroit Diesel to 440hp (298.2kw)
Transmission: Mack Maxitorque five speed or Fuller 10-speed transmission
Payload: not available
Applications: long-distance, on-highway, severe-service off-highway
Special features: designed extra strong for heavy-hauling applications.

The Super-Liner was put to work in a wide variety of heavy-duty applications, hauling everything from livestock to tankers. Nearly 20,000 RW Super-Liners were built before the model was retired.

Engine options

The Super-Liner was designed to accommodate high-horsepower engines and the large radiators require to cool them. In fact, a wide range of power options were available, including the new line of Mack fuel-efficient engines, such as the Caterpillar 3406 with up to 400hp (298.2kw), the Cummins Formula 300, and the Detroit Diesel 92 Series, which could deliver up to 435hp (324.3kw). The Mack option was popular because it delivered weight savings of up to 600lb (272kg) and the strength of 440 horses. The extra strength and reduced weight was partially be attributed to the new chassis-mounted charge-air cooling system.

Quiet ride

Although the Super-Liner was far from aerodynamic-looking by today's standards, the designers were still cognizant of the importance of air flow. They built the truck with no exterior protrusions so air would flow smoothly over the chassis, providing a quiet ride. The double-wall cab was built of steel and paired with a unitized fiberglass hood-fender assembly, which saved weight while providing extra strength.

Foden

chapter 8

New Technologies

In the 1980s and 1990s new technology would have a huge impact on the trucking industry. Trucks became safer, more powerful, and much more environmentally friendly. In Europe operators were encouraged by some manufacturers such as Scania to adopt a more modular approach to truck purchasing by using common components to reduce overheads.

Above: *The Mack CH was described at the time of release as one of the most versatile trucks on the road.*

Left: *The Alpha reestablished Foden as a major player in the heavy-duty truck market.*

There were plenty of changes to the trucking industry in the U.S. during this era as well. In 1982 legislators began relaxing length restrictions and replacing them with weight restrictions. The change meant cab-over-engines were no longer required to maximize payloads and conventional-style trucks once again increased in popularity. Cab-over-engine trucks have all but disappeared from the highways of North America since then.

Europe may have been leading the way when it came to aerodynamic styling, but the U.S. wasn't far behind and by the mid-1980s American trucks were incorporating many of the aerodynamic design elements pioneered overseas. Other innovations followed: Detroit Diesel became the first U.S. engine manufacturer to offer electronic fuel management on its Series 60 engines. In terms of safety, antilock braking systems (ABS) made their debut on trucks in 1986. Mercedes-Benz and Scania equipped their trucks with disk brakes to reduce the stopping distances of their vehicles in 1998 and that same year Volvo began offering airbags on the driver's side. The increased use of air suspension systems and greatly improved roads at this time also offered unprecedented levels of driver comfort.

SCANIA

1980 Sweden

PROGRAM SCANIA

At the beginning of the 1980s, Scania developed a new line of trucks, Program Scania or the GPRT range (a.k.a. 2-series), to meet the needs of driver comfort, reliability, good fuel consumption, and efficient service.

A modular philosophy drove the 2-series. Scania felt that by using common components, it could realize more economical production of vehicles while building an unlimited number of truck variants. Stig Ericsson was the engineer credited with introducing the modular philosophy to Scania. Work had begun behind the scenes as early as 1974, when planners began studying existing and future demands in different market segments. Ericsson was responsible for chassis design and production. Not only did the company modularize its engines, gearboxes, drive shafts, and final gears, but it took it one step further and included chassis components such as axles, frames, and cabs.

Using this modular approach, Scania developed a range of vehicles that offered a wide variety of trucks in the 15.7 to 38.6-ton (16 to 38-tonne) GVW range. Although the 2-series trucks shared many components, there was still plenty of room for customization under this system.

Improved efficiencies

Program Scania resulted in the replacement of three different cab ranges with a single, modular range. While customers could still choose the options they wanted, the number of components in the original cab range was reduced by 70 percent. The amount of sheet metal required was also dramatically reduced and the company became more efficient, reducing the average working hours per cab by about 30 percent.

The end-users embraced the concept, and when the 2-series was replaced in 1988 more than 170,150 of these trucks had been sold in 116 versions. One of the most significant benefits for trucking companies was the fact the cost of building these trucks was significantly reduced so savings could be passed onto the customer.

Above: *A full range of Program Scania trucks made up the 2-series. Pictured here from left to right is the 82M, 112E, and the 142H.*

Above, top: *A single modular cab replaced three different cab types and a hooded "T" cab version also shared certain components.*

Specifications

Country: Sweden

Year manufactured from: 1980

Engine: turbo 475 or 674ci (7.8 or 11 liters)

Transmission: five- or 10-speed

Payload: GVWs of up to 35.4 tons (36 tonnes)

Applications: long-distance, on-highway

Special features: common components used to increase efficiencies.

VOLVO

1980 Sweden

VOLVO CH230

Volvo Trucks was already a major player in the global truck market by the late 1970s, with most of its mass-produced vehicles performing a wide range of applications throughout the world.

Right: *Volvo introduced the CH230 mainly for Swiss carriers that had to operate slightly narrower trucks and trailers. The reason was that many Swiss highways were narrow and windy, making them dangerous for wide trucks to travel.*

Above: *It wasn't uncommon for the CH230 to feature four axles because of Switzerland's strict weight limitations.*

Specifications

Country: Sweden
Year manufactured from: 1980
Engine: not available
Transmission: eight or 16-speed
Payload: not available
Applications: long-distance, on-highway
Special features: narrower truck designed for Swiss market.

Since Switzerland often faced harsh winter weather and had a challenging terrain for truckers to navigate, the country restricted the width of commercial vehicles to 7ft 6in (2.3m). Most other European countries, meanwhile, allowed vehicle widths of up to 8ft 2in (2.5m). Some observers also suggested the Swiss legislation was aimed at reducing the efficiency of trucks to encourage shippers to move their products by rail. Whatever the reason, Volvo's new FH10/FH12 would not be permitted in Switzerland, since designers were intent on making that vehicle a full-width truck to provide a more spacious cab and longer bunk.

Powerful, yet narrow

Rather than leave its Swiss customers high and dry, Volvo developed a special CH230 model exclusively for its Swiss customers. This combined elements of the F89 with the narrow front axle of the F86 and a narrow N10 rear axle, joining the F89's power with the narrower axles of other Volvo trucks and creating a vehicle of tremendous power. However, that version was only tentative. An entirely new truck was still in development and this new CH230 was introduced at the Geneva Truck Show in 1980. The new CH230's chassis was taken from the F10/F12 but included the narrow axles of the F7.

KENWORTH
🛠 **1981 USA**

KENWORTH C500

Kenworth introduced its heavy-hitting C500 to the industry in 1981. It has gone on to establish itself as one of the strongest, most hard-wearing trucks in the market and is called on for extremely heavy hauls.

It was introduced in 6x4 and 6x6 configurations powered by a range of Caterpillar, Cummins, and Detroit Diesel engines. Typically, early versions of the C500 would be paired with a six-cylinder 270hp (201.3kw) Cummins NTC270 powerplant. Other common specifications included a Fuller RTO1157DL 8F 1R transmission and Eaton single-reduction spiral bevel tandem rear axles.

Today's C500

Today, Kenworth refers to the C500 as its serious workhorse truck. It has been used to haul loads of more than 196.8 tons (200 tonnes) of coal through the jungles of Southeast Asia and to move oil rigs through Alberta muskeg. It is built to take on the toughest tasks whether they at the construction site or in the wilderness.

The C500 family features heavy-duty frame rails to provide maximum carrying capacities. Front axles rated up to 30,000lb (13,605.4kg) and were available in both driving and nondriving configurations. Meanwhile, carrier-reduction rear tandem axles rated up to 70,000lb (31,746kg) also contribute to the truck's strength. Since the truck is often operated off-road, there is optional air suspension.

Above: *Kenworth built the C500 specifically for the demands of the construction industry.*

Specifications

Country: USA
Year manufactured from: 1981
Engine: full range Cat, Cummins and Detroit Diesel
Transmission: Fuller RTO1157DL 8F
Payload: not available
Applications: construction, severe-duty, off-highway
Special features: used to haul up to 196.8 tons (200 tonnes) of coal through Asian jungle.

MACK

1983 USA

MACK ULTRA-LINER

Above: *Mack's reputation for the strength and durability of its trucks was reinforced with the introduction of the Ultra-Liner. A glass-polyester skin covered a steel frame resulting in improved corrosion resistance and durability.*

When Mack introduced its MH Ultra-Liner series in 1983, the company billed it as "the truck bred for the '80s and beyond." The cab-over-engine truck premiered the new Maxi-Glass composite cab.

Right: *Mack's Ultra-Liner was the first truck to utilize the company's Maxi-Glass composite cab. Notice the slightly sloped cab, which helped improve the aerodynamics of the truck.*

Specifications

Country: USA
Year manufactured from: 1983
Engine: not available
Transmission: five or 10-speeds to bevel
Payload: up to 19.6 tons (20 tonnes)
Applications: long-distance, on-highway
Special features: slightly sloped body provided better aerodynamics.

For one thing, it was aerodynamic, which may seem like an oxymoron when referring to a cab-over-engine vehicle. But, nonetheless, it had a slightly sloped body that provided better airflow. Comparisons have been made between the later model MH Ultra-Liners and the CL cab-over trucks offered by Ford. The MH Ultra-Liner featured a massive grille and rectangular headlights located inside its frame.

Popular in the east

In addition, it was extremely strong, being comprised of GRP built around a steel frame, which also made it more corrosion-resistant than traditional truck cabs. The MH Ultra-Liner could be customized as either a sleeper cab or a day cab, depending on the needs of the user.

Mack discontinued production of cab-over-engine trucks altogether in 1994 as length restrictions were loosened in the U.S. and the popularity of conventional-styled trucks won out. At around the same time, Renault doubled its investment in Mack, purchasing 20 percent of its shares. Mack began marketing Renault's cab-over Mid-Liner city truck in North America. The Mid-Liner series delivered excellent drivability and payload for urban applications.

MITSUBISHI

1983 Japan

MITSUBISHI FUSO "THE GREAT"

Ten years after the F series established itself as a force for Mitsubishi, the line was completely redesigned, both inside and out. The new-look F series introduced in 1983 was dubbed "The Great" in Japan.

Above: *The Mitsubishi Fuso "The Great" was powered by a peppy, yet fuel-efficient, diesel engine that made the truck ideal for hauling heavy loads over long distances.*

Some of the enhancements included a new aerodynamically designed cab as well as a larger windshield for increased visibility. In fact, the new windshield was the largest in its class in Japan. A full floating cab suspension also provided for a better ride and less wear and tear on components. The 6D22T2 diesel engine and its intercooler and turbocharger provided plenty of power for "The Great" new Mitsubishi. This engine also boasted improved fuel mileage and torque characteristics making it extremely efficient.

A seven-speed transmission with overdrive was available on some new models. Many of them also featured the new MCVOCII on-board computer, which provided information about operating parameters required for proper servicing and maintenance.

Plenty of enhancements

Development didn't end with the launch of "The Great" in 1983, however. Five years later Mitsubishi introduced further improvements. One example was the world's first prestroke control injection pump for diesel engines, introduced in 1988. Mitsubishi also introduced a finger-control transmission and the MMAT automatic transmission. The following year, Mitsubishi integrated new ABS technology into its trucks, improving safety. In 1991 the company developed the "super frame," which provided simplified rear-body mounting combined with higher rigidity and a lower overall weight. Once again, these enhancements improved the efficiency and economy of the truck considerably.

Specifications

Country: Japan
Year manufactured from: 1983
Engine: 6D22T2 diesel engine
Transmission: seven-speed transmission with overdrive
Payload: not available
Applications: local pickup and delivery
Special features: full floating cab suspension provides smooth ride.

DAIMLER-BENZ
1984 Germany

DAIMLER-BENZ LK (LN2)

Daimler-Benz overhauled its industry-leading lightweight LP models in 1984, replacing them with the new line of trucks known as the Light Class (LK) or LN2. This model never in fact received an official name.

Above: *The LK series was available in a wide range of sizes and configurations making it a very diverse truck.*

Specifications

Country: Germany
Year manufactured from: 1984
Engine: OM 364 and OM 366 with up to 204hp (152.1kw)
Transmission: five-speeds to bevel
Payload: 8.13 tons (eight tonnes)
Applications: local pickup and delivery, some long-distance
Special features: tilt cabs provided easier access to the engine and other components.

Among the enhancements to the new light trucks were tilt cabs for improved access to the engine and other components, as well as improved styling.This characteristic reflected the Bremen van design made famous in 1977.

In addition, the Light Class vehicles were available in a wider range of sizes and configurations ranging from 6.3 to 12.7 tons (6.5 to 13 tonnes) and featured engines ranging from 90 to 204hp (67.1 to 152.1kw). A long-distance cab was available, making this medium-duty offering more versatile than many others on the market at that time. The Light Class trucks were also the first commercial vehicles to be equipped with low-profile tires.

Sophisticated engines

The origins of the OM 364 and OM 366 engines lay in the OM 312 engine series introduced in 1949. However, these later engines were far more advanced. The OM 364 was a four-cylinder engine while the OM 366 had six cylinders. While the older OM 312 may have provided the inspiration for the new engines, a great deal of sophisticated engineering had been incorporated.

In 1996, Daimler-Benz introduced the upgraded Light Class vehicles. The most notable improvement was the new 900-series ECO POWER engines. These new powerplants were turbocharged in-line four- and six cylinder units with intercooling.

KENWORTH

1985 USA

KENWORTH T600A/T800

Kenworth introduced the T600A in 1985, a revolutionary sloped-nose conventional tractor with a front axle that was set back. The result was the comfort of a conventional and the maneuverability of a cab-over.

At first, truckers were slow to embrace the radical new design. American truckers have traditionally enjoyed the long, square, conventional-style hoods Kenworth and Peterbilt were known for. The new-look T600A took some getting used to. Money talks, however, and with diesel prices reaching new heights it wasn't long before truckers were willing to give it a shot.

Massive fuel savings

The T600A was more aerodynamic than its predecessors, which helped customers shave up to 22 percent off their fuel costs. Compared with Kenworth's W900 conventional, the T600A delivered aerodynamics that were 40 percent better. The company admitted it had originally feared the radical new styling might turn off some customers. In the end, the gamble paid off because it has earned its spot at the top of Kenworth's best-seller list.

Still basking in the success of the T600A, Kenworth launched the T800 a year later. This truck featured a set-back front axle for maximum payload and maneuverability but

Above: *The T600A and the T800 (seen here) accounted for half of Kenworth's production by the mid 1990s.*

it was designed for heavier-duty operations. It was also suitable for off-highway applications such as logging and mining. The T600 and T800 have been frequently updated since their introduction. The two models accounted for half of Kenworth's production by the mid-1990s. While engineers continue to find ways of saving even more fuel with the T600 and T800 series, unfortunately for truckers, the cost of diesel seems to be keeping up with any enhancements.

Today's T600

Kenworth insists its T600 is without equal when it comes to aerodynamics, performance, and driver comfort. It is an ideal truck for linehaul operations (movement of products between two customer locations, for example, a local warehouse and a regional distribution center) thanks to its exceptional aerodynamics, and its set-back front axle provides excellent maneuverability and weight distribution.

The truck can be customized for a range of applications including regional pickups and deliveries. Bulk haulers demanding high payload capacity can also trim it out through a variety of weight-saving options. One could say it's the truck for every job.

Drivers appreciate its quiet cab and conveniently located gauges and controls. It also provides a smooth ride because of its 64in (162.5cm) front springs, a proprietary eight-bag air suspension as well as a cab and/or sleeper suspension. To all this is added the fact that the T600 is known for its ease of maintenance and high resale values and it's no wonder the model has proved to be one of Kenworth's most successful.

Today's T600 can be matched with a wide number of engines ranging between 610 and 976ci (10 and 16 liters) and up to 600hp (447.4kw). The front axles are available in ratings between 12,000 and 14,600lb (5,454 and 6,636kg) while the rear axles are rated at 23,000lb (10,455kg) for singles and 46,000lb (20,909.6kg) for tandems. The sleepers that are available on the T600 line range from the 38in (96.5cm) AeroCab FlatTop to the more spacious 86in (2.18m) Studio AeroCab.

Left: *The T600A was an immediate hit with long-distance haulers where fuel mileage was a major concern. The sloped nose remains popular today.*

Specifications

Country: USA

Year manufactured from: 1985

Engine: full range up to 600hp (447.4kw)

Transmission: nine or 15-speeds to bevel

Payload: not available

Applications: long-distance, on-highway

Special features: sloped hood claimed to reduce fuel consumption by 22 percent.

Below: *The aerodynamic T600A was a radical departure from Kenworth's traditional long-nose conventional-styled truck. While it took some getting used to, even Kenworth loyalists couldn't argue with a 22 percent fuel savings.*

IVECO

1984 Italy

IVECO TURBOSTAR

Iveco was founded in the mid-1970s, but it struggled to produce trucks that reflected its own corporate identity. This could be partly attributed to the fact that it was comprised of various factories and companies.

Specifications

Country: Italy

Year manufactured from: 1984

Engine: Iveco 8210 842ci (13.8-liter), 330hp (246kw), or the Iveco 8280 1,049ci (17.2-liter) with 420hp (313.1kw)

Transmission: eight-speeds to bevel

Payload: not available

Applications: heavy-duty, long-distance

Special features: TurboStar was Iveco's first success in heavy-duty truck market.

However, in 1984 Iveco introduced the TurboStar, which attracted some attention in the trucking industry and helped establish the company as a real player in the commercial-vehicle industry. The TurboStar was a huge success in the heavy-duty truck sector. The truck's popularity could be attributed in large part to the cab and engine.

The cab looked similar to those produced by FIAT, but it included many enhancements including an interior height of 5ft 7in (1.7m). The cab was more comfortable than other common cabs of the day and a softer suspension system also contributed to the TurboStar's popularity because it delivered a smoother ride. Iveco referred to the truck as the first real blend of different cultures, which had been the focus of Iveco's efforts right from the start. The Iveco TurboStar became famous as the car-hauler of choice for the Ferrari Formula 1 team.

Most powerful engine available

The TurboStar could be purchased with one of two engines. They included the 8210 842ci (13.8-liter) straight-six turbo with 330hp (246kw), or the 8280 1,049ci (17.2-liter) behemoth with 420hp (313.1kw). The latter, a V8, was one of the most powerful engines available in any truck at the time making the TurboStar an ideal truck for heavy-duty haulage. It was particularly popular for long-distance and international transportation.

Above: *The Turbostar was Iveco's first real success in the heavy-duty long-distance trucking sector. An interior height of 5ft 6in (1.7 metres) made it an ideal truck for long-distance haulage.*

SCANIA 3-SERIES
1988 Sweden

Above: *While the Scania 3-series closely resembled its 2-series predecessor, it incorporated numerous refinements to improve performance and reduce emissions.*

SCANIA 3-SERIES

Scania introduced its 3-series in 1988 as a successor to the popular 2-series of heavy-duty trucks. Most of the developments incorporated into the 3-series revolved around the powertrain.

Right: *Scania's 3-series was highly customizable and available in a wide range of configurations. Its reliability helped win it European Truck of the Year honors in 1989.*

Specifications

Country: Sweden
Year manufactured from: 1988
Engine: full range
Transmission: five or 10-speeds to bevel
Payload: not available
Applications: long-distance, on-highway
Special features: winner of the European Truck of the Year award in 1989.

However, improvements were by no means limited just to the drivetrain components. The 2-series' dashboard was replaced with a new, more ergonomic type that curved toward the driver, providing easy access to important gauges and instruments. The driver also enjoyed a smoother ride thanks to a cab with four-point suspension.

The Scania 3-series was very well received by the trucking industry and it took top honors as Truck of the Year in Europe in 1989. It was the first time an entire truck range had won the award. The Truck of the Year jury commented that the 3-series boasted: "A generally high standard of engineering, top-class driving comfort combined with a high standard of driver and passenger safety, advanced engines offering good fuel consumption, and superb reliability."

Modular design

With the development of the 3-series, Scania was forging ahead with its modular approach. The manufacturer was producing "made-to-measure" trucks for specific customers, depending on their individual needs. Customers were able to specify exactly which components they wanted on their vehicle.

DAIMLER-BENZ

1988 Germany

DAIMLER-BENZ SK

Daimler-Benz's NG (new generation) resulted in the birth of the heavy-duty (SK) class in 1988. It was designed to reflect the styling of the popular NG trucks and borrowed heavily from the NG's outward appearance.

Above: *Daimler-Benz introduced Low Emission Vehicle (LEV) engines in 1991, reducing pollutant emissions by 50 percent.*

Specifications

Country: Germany

Year manufactured from: 1988

Engine: Mercedes-Benz V6, V8, or V10

Transmission: 12-speeds to bevel

Payload: not available

Applications: long-distance, on-highway

Special features: when introduced the SK was the most powerful highway truck available in Europe.

Above, left: *The famous Mercedes-Benz badge dates from 1926 when Daimler merged with Benz. The Daimler three-pointed star was combined with the circular laurel garland of Benz.*

The SK trucks did establish their own identity, however, and could be recognized by their new windows, an angled cowling, and a broad radiator grille.

Incredible power, and braking power

The cab was completely redesigned to provide a better driver environment. And the thoroughly reengineered engines capable of delivering 260, 290, 354, 435, or 475hp (193.8, 216.2, 263.9, 324.3, or 354.2kw) were able to deliver top performance. The 475hp (354.2kw) version was the most powerful highway truck available in all of Europe at the time, Daimler-Benz proudly claimed. By 1994 the 400 series engines were capable of delivering 530hp (395.2kw).

Another huge development in the SK trucks was the improved brake performance. The engine brake featured a constantly open throttle valve, which greatly increased the engine's braking performance. This technology was later renamed the "decompression valve engine brake."

The demand for cleaner engines was continuing to be an issue in Europe at the time and the SK powerplants made significant advances in this area. In 1991 Daimler-Benz introduced its Low Emission Vehicle (LEV) engines, reducing pollutants and particulate emissions by more than 50 percent. This reduction in emissions can also be attributed in part to the company's Electronic Diesel Control (EDC) system.

MACK
1988 USA

MACK CH

Mack Trucks introduced its CH highway lineup to the North American trucking industry in 1988. The CH was an axle-forward vehicle suitable for on- and off-highway stop-and-go applications.

Right: *Mack referred to its CH series as one of the most versatile trucks on the road. It was put to the test in many applications both on- and off-highway.*

Specifications

Country: USA
Year manufactured from: 1988
Engine: Mack ASET
Transmission: 10-speeds to bevel
Payload: not available
Applications: heavy-duty on- and off-highway, vocational
Special features: set-forward axle provides excellent maneuverability.

A CH day cab was included in the line, which was ideal for lightweight bulk haul, van, flatbed, and lowboy applications, the company said. At the time of its release, Mack said the CH was one of the most versatile trucks on the road because it could be customized for a wide variety of heavy-duty highway and vocational applications. A range of horsepower ratings were available, which could be paired with various transmissions to deliver optimum performance.

When designing the CH cab, engineers set out to combine the best ergonomic styling with traditional heavy-truck features. For instance, switches and gauges were placed within easy reach of the driver while the dashboard was well-lit, making it more visible. The cab was mounted on an air-ride suspension to give a smooth ride. The driver's view was enhanced by the sloped hood.

Three sleepers

Mack provided three types of sleeper box for the CH series including 48in, 56in, and 70in (1.21m, 1.42m, and 1.77m) options. Engineers utilized every inch of the cab's space by providing plenty of storage in areas such as the space under the bunk. Improved insulation within the sleeper box provided a more comfortable environment.

VOLVO
1989 Sweden

VOLVO NL SERIES

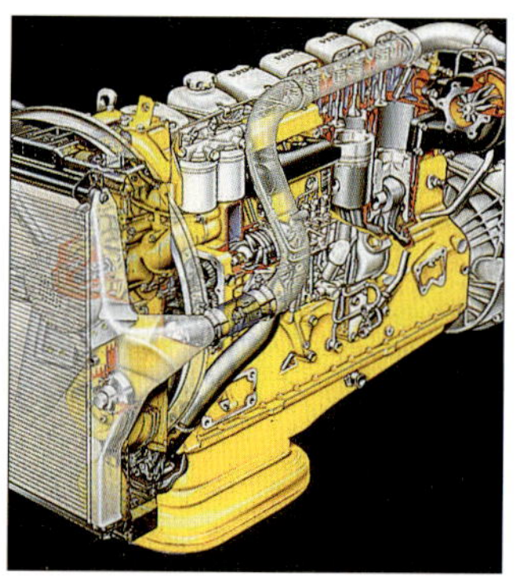

Above: *Volvo NL customers could choose either a 607 and 728.4ci (10- or 12-liter) engine. Both were known for their efficiency.*

Most European truck manufacturers focused all their attention on building cab-over-engine trucks, but Swedish manufacturers were the exception. Volvo, for instance, offered a variety of conventional trucks.

Specifications

Country: Sweden
Year manufactured from: 1989
Engine: 607 and 728.4ci (10- and 12-liter) Volvo engines
Transmission: eight or 16-speeds to bevel
Payload: not available
Applications: long-distance, on-highway
Special features: truck was designed for Brazilian market.

Above: *While cab-over-engines were the trucks of choice in Europe, Swedish manufacturer Volvo was building conventional-style trucks. The NL became a very popular truck in Brazil where it soon replaced previous Volvo conventional-style trucks.*

The N-type was continuously improved on to deliver more spacious cabs and more economical powerplants. Like their F-type sisters, these trucks also featured advances when it came to transmissions and chassis components. Most of the N trucks could be customized with either a day cab, sleeper cab, or crew cab making it a fairly versatile vehicle.

The first of Volvo's N trucks, the N7, N10, and N12, were introduced in 1973. They became a hot commodity in Brazil and other overseas markets. The company introduced the NL to Brazil in 1989 and it was seen as an improvement on existing N trucks.

Roomy interior

The NL10 and NL12 were more aerodynamic than previous N-type trucks offered by Volvo and some of the aerodynamic styling of the truck was borrowed from Volvo's North American truck line.

Brazilian drivers loved the new truck. Its long hood provided for a more spacious interior and better access to the other side of the cab. Drivers also enjoyed the fact the cab's interior was better insulated from the engine compartment. The NL10 and NL12 became so popular in Brazil that they soon replaced all other existing Volvo N-type trucks.

KENWORTH
1989 USA

KENWORTH W 900L

At the end of 1989, Kenworth launched the W 900L (L for "Long"), with an extended hood. It was a special variant of the great W 900, which is still manufactured after more than four decades.

Right: *The Kenworth W 900L remains by far the most stylish American truck in the world, with a massive square radiator and an external stainless air filter.*

Specifications

Country: USA
Year manufactured from: 1989
Engine: Caterpillar or Cummins up to 600hp (447kw)
Transmission: full range
Payload: up to 200,000 GCW, normally 80,000 GCW
Applications: long-distance, on-highway
Special features: extended hood 130in (330cm) BBC, engines up to 600 hp (447kw).

With the introduction of new big six-inline turbocharged and aftercooled engines from Cummins and Caterpillar (up to 600hp/447kw), Kenworth had to change some details of the famous W 900: an extended hood and bigger radiator were now necessary. The result was the new W 900L. The tilted fiberglass hood was so long that it was gently sloped to improve the visibility.

Classic styling

Of course, the W 900L kept the classic styling of its predecessors, with plenty of chrome, decorative stainless-steel parts, and custom upholsteries. There are now 19 interior color options and three main trim levels (Splendor, Diamond, and Diamond VIT—Very Important Truck). For comfort, the handcrafted Studio Sleeper and the Aerocab/Aerodyne Sleepers give the riders plenty of room and a much more of a just-like-home feel. In order to save weight, the sleepers and the cab are made of corrosion-proof aluminum and fiberglass materials. A modern air-suspension system provides a great smooth ride.

With a length of 130in (330cm) bumper to back of cabin (BBC), the W 900L is specially designed for long-haul applications and hill climbing and for owner-operators. It has a high resale value, and the W 900L remains the spearhead of Kenworth, even if it's not the best seller of the company since the introduction of the T 600 and the T 800.

IVECO

1990s Italy

THE IVECO EURO RANGE

In the early 1990s Iveco introduced the Euro range that included trucks with gross vehicle weights ranging from 5.9 tons (6 tonnes) to the maximum legal limit (depending on the market).

The Euro range was immediately embraced by the industry. The EuroCargo and the EuroTech won the prestigious Truck of the Year award in 1992 and 1993 respectively. It marked the first time in the award's 17-year history that the same manufacturer won in consecutive years.

The EuroStar was customizable for various applications, such as B-double, highway, or roadtrain, and capable of gross vehicle weights of up to 88.5 tons (90 tonnes). It was marketed as an over-the-road truck and was powered by straight-four, straight-six and V8 engines ranging from 238 to 1048ci (3.9 to 17.2-litres) and giving power between 102 and 512hp (76 and 383kw) Like American trucks, the EuroStar featured components from a variety of suppliers including Eaton and Meritor.

Engine options

The EuroTech was also capable of handling a variety of applications including local delivery, highway, dump, B-double, and road train. Its power could be supplied by an Iveco, Cummins, or Detroit Diesel engine. The EuroTrakker remains a popular truck for construction and other off-road applications. It is one of the most technologically advanced trucks in the construction sector and boasts extremely high torque outputs even at low speeds. The cab panels are galvanized on both sides, providing excellent soundproofing and driver protection

Above, top and left: *The new Iveco Multi-Purpose or "MP" cab gave the range a new corporate identity.*

Specifications

Country: Italy

Year manufactured from: 1990

Engine: Iveco 238 to 1049ci (3.9- to 17.2-litres)

Transmission: 12-speeds to bevel

Payload: 2.95 to 23.6 tons (3 to 24 tonnes) dependant on model

Applications: local delivery, highway, dump, B-double, and road train

Special features: famous Italian design company, Stil Bertone, contributed to the design of the EuroCargo cab.

MITSUBISHI

1993 Japan

MITSUBISHI FUSO SIXTH-GENERATION CANTER

Mitsubishi named its latest version of the Fuso "The Super Great," and when it launched its sixth generation Canter, it dubbed it simply a "Good Truck." This was the fifth model change in the Canter's 20-year history.

Specifications

Country: Japan

Year manufactured from: 1993

Engine: 140hp (104.3kw) Mitsubishi engine

Transmission: five-speeds to bevel

Payload: two to 3.94 tons (two to four tonnes)

Applications: local pickup and delivery

Special features: liquid-filled cab suspension provided smoother ride.

Above: *The sixth generation Canter featured a rounded cab for improved aerodynamics. The redesigned headlights also provided better nighttime visbility.*

The launch in 1993 of the sixth-generation Canter came during the truck's 20th anniversary year. The first noticeable difference was be the rounded-cube cab with improved aerodynamics. The new Canters were also paired with high-performance engines including the new 140hp (104.3kw) naturally-aspirated direct-injection engine with the highest power in its class at the time.

Mitsubishi had redesigned the Canter to provide the best in productivity and comfort for users of all types and in order to achieve this, the manufacturer introduced many new enhancements. "Projector" headlights were among the safety enhancements introduced on the sixth-generation Canter, providing better visibility at night.

Liquid-filled suspension

A liquid-filled cab suspension was another first to be introduced on the sixth-generation Canter. The new truck also had a front wheel independent suspension giving it the feel of a luxury car, as opposed to a typical delivery truck. The new Canter's interior was esthetically pleasing and featured increased comfort and improved ergonomics.

VOLVO
1993 Sweden

VOLVO FH SERIES

Volvo's FH series, introduced in 1993, has been a true success story for the Swedish manufacturer. The FH12 and FH16 have been heralded as among the most popular trucks ever to take to the highway.

One of the factors contributing to the FH series' success was its cutting-edge technology, which helped improve areas such as fuel efficiency. The FH12 was powered by an all-new 732ci (12-liter) D12A engine engineered by Volvo. The FH16, meanwhile, featured a revamped Volvo 976ci (16-liter) engine that was first introduced in 1987 under the hood of the F16.

The D12A was designed for use on both sides of the Atlantic. It was the first European-designed diesel truck engine to boast a high-pressure unit fuel injector and overhead camshaft along with four valves per cylinder. This combination provided better breathing for the engine, which led to higher efficiency and lower fuel consumption. This technology was previously only found in advanced passenger car engines. In addition, it helped reduce emissions, which was becoming more of an issue in the 1990s.

The D12A also featured Volvo's own engine brake, which contributed to the safety of the new engine. The Volvo Engine

Above: *The FH16 looked much like its less powerful cousin, the FH12, but a 976ci (16-liter) engine provided considerably more power.*

Brake was necessitated by the engine's ability to maintain higher average speeds, which caused more strain on the brakes. Volvo's Accident Commission—a team that has investigated accidents for more than 25 years—also lent its expertise to the designers to ensure the FH series was as safe as possible.

Aerodynamic cab

The FH12 and FH16 featured cabs that were designed with the driver in mind. Fuel efficiency was also a motivation for engineers, and they designed the cab to be aerodynamic and lightweight without sacrificing any of the comforts drivers expected and enjoyed. In fact, while making the cab more efficient and productive, designers also made it quieter and more comfortable. Three cab versions were introduced for the FH12 and FH16. They included a short day cab, long sleeper cab, and the Globetrotter, a long sleeper cab with interior standing room. Two years later the even larger Globetrotter XL was introduced.

More enhancements introduced

The FH became the first heavy-duty truck to be equipped with a driver's side airbag in 1995, according to Volvo. This introduction elevated the truck's passive safety to a whole new level.

In 1998 the FH vehicles received a facelift and the result was the second Truck of the Year Award for the lineup, which was presented in 2000. The FH series received its first such accolade in 1994 following its introduction. The look of the truck changed somewhat as a result of the enhancements. The front of the cab was altered to improve airflow and increase the efficiency of the headlights. Another safety feature was introduced along with the new FH series at this time—Volvo's new electronically controlled heavy-duty disk brakes.

Another enhancement introduced in the late 1990s was the new D12C engine. It had been completely retooled to provide better performance and greater fuel economy as well as extended service life and even better environmental characteristics. The new engine was developed with impending Euro 3 emissions standards in mind, which were to come into play in 2001.

Volvo halted production of the FH16 in 2001, but in 2003 it came out with a completely new 976ci (16-liter) in-line six cylinder direct-injection diesel engine with overhead camshaft, four valves per cylinder, and electronic unit fuel injectors.

Left: *Many truck enthusiasts insist Volvo's FH12 was the most successful truck to ever be built. The critics agreed, naming the FH-series Volvo the Truck of the Year in 1994 and 2000, the first time a single truck won the award twice.*

Specifications

Country: Sweden

Year manufactured from: 1993

Engine: Volvo D12 A or D16A

Transmission: nine or 14-speeds to bevel

Payload: 9.8 to 39.4 tons (10 to 40 tonnes)

Applications: long-distance, on-highway, logging

Special features: Volvo Engine Brake provided stopping power to reduce strain on foundation brakes.

Left: *The FH12 could be purchased with the Globetrotter cab, which was big enough to allow a driver to stand up inside. An even larger Globetrotter XL cab was introduced soon after.*

ERF
1993 **Holland**

ERF EC

ERF was one of the first manufacturers to consult drivers while designing a new truck, and it showed when it unveiled its EC range in 1993. Some of the feedback resulted in a raised bunk and a wraparound dashboard.

Other features that drivers were quick to embrace were electronically adjustable mirrors and extra storage space underneath the raised bunk. The cab was extra spacious, ideal for long-distance truckers.

ERF's EC range consisted of the EC8, EC10, EC12, and EC14, with the EC10 being the most common. It was powered by a Cummins 607ci (10-liter) six-cylinder Euro-1 compliant engine with power ratings of up to 350hp (260.9kw). Within a couple of years, more powerful engines were available and the EC14 was capable of 525hp (391.4kw). At this time, engines had to comply with Euro-2 emissions standards.

While the drivers liked the many enhancements in the cab, owners were impressed by the truck's fuel efficiency, which resulted from improved fuel injection and combustion systems.

Continuously updated

ERF continued to survey its customers to see what enhancements drivers would like to see in future versions of the EC range. As a result of this process, the dash was lowered to enhance visibility and a more spacious footwell was created. Better heating and seating were also improvements that worked their way into the truck line over time.

Above: *The ERF EC 14 engine had up to 525 hp (391.4kw) making it ideal for heavy haulage.*

Specifications

Country: United Kingdom
Year manufactured from: 1993
Engine: Cummins 488ci or 854ci (8 or 14-liter)
Transmission: nine or 12-speeds to bevel
Payload: not available
Applications: long-distance, on-highway
Special features: among the most fuel-efficient trucks in the industry.

ASHOK
1994 **India**

ASHOK LEYLAND CARGO

Above: *The Ashok Leyland Cargo was used to haul a variety of freight—some of it rather unusual, such as this elephant, for instance.*

One of Ashok Leyland's most successful launches has been the Cargo. The Cargo brought a new set of values into the Indian trucking industry.

Right: *The truck was also used in more common applications as dump trucks, garbage trucks, and even fire trucks.*

Specifications

Country: India
Year manufactured from: 1994
Engine: naturally aspirated Iveco diesel
Transmission: five-speed synchomesh transmission
Payload: not available
Applications: pick-up and delivery, refuse
Special features: environmentally friendly medium-duty truck.

The Cargo was used for a wide range of applications, particularly for refuse collection or the delivery of general freight. It was also equipped as a fire truck by some customers. It was commonly matched with a naturally aspirated Iveco diesel engine after Ashok was bought from Leyland.

New Values

The truck was designed to be technologically advanced as well as ecofriendly. Driver comfort and safety and vehicle reliability were all areas the manufacturer was determined to set a new standard for with the cargo. Ashok Leyland had the experience. It was already a manufacturer of world-leading specialty vehicles and engines and it was also the first company to introduce multiaxle trucks to the Indian trucking industry. It had key partnerships that would allow it to be successful, including an important one with Iveco. Iveco's commercial vehicle technology was used in the development of the Cargo. Still, a $200 million investment was needed to begin production of the Cargo range.

KENWORTH
1994 USA

KENWORTH T300

Kenworth launched itself into the medium-duty conventional market in 1994 with the introduction of the T300. The T300 medium-duty was marketed toward customers looking for a premium truck.

Above: *Kenworth entered the medium-duty market in 1994 with the T300.*

The T300 Class 7 truck featured a modified T600 cab and had a standard gross vehicle weight of 30,000lb (13,605.4kg) although higher axle ratings were available. The T300 tractor, meanwhile, had a 65,000lb (29,478.4kg) gross combination vehicle weight rating.

Custom engineered

The truck was built to be tough, with features including long-lasting huckbolt fasteners. It was also built using readily available components to make maintenance easier and less expensive. The T300 was custom engineered to match specific applications in the medium-duty market segment. Some of those applications the T300 was designed for included pickup and delivery, fuel oil, propane, mixer, dump, refuse, wrecker, expedited freight, fire and rescue, and refrigerated food products.

The goal of the vehicle was to deliver superior return on investment and the truck was built to meet Kenworth's reputation as a builder of premium Class 8 vehicles. Maneuverability, visibility, and versatility were a few of the selling points for the Kenworth T300. It featured a 20° sloped hood and 50° wheel cut to help meet these targets. A one-piece windshield also helped improve visibility while the Daylite doors allowed a driver to see the ground through windows on the lower part of the doors. West Coast mirrors were mounted to the cab cowl for increased durability and the need for fewer adjustments. The cab was comprised of aluminum and GRP provided excellent strength and durability.

Specifications

Country: USA
Year manufactured from: 1994
Engine: full range of Cummins and Cat engines up to 330hp (246.1kw)
Transmission: full range of mechanical and automated transmissions.
Payload: eight to 18 tons (8.1 to 18.3 tonnes)
Applications: pickup and delivery, fuel, propane, mixer, dump, refuse, wrecker, expedited freight, fire, and rescue.
Special features: Daylite doors improve driver visibility to the sides.

VOLVO

1995 Sweden

VOLVO ENVIRONMENTAL CONCEPT TRUCK

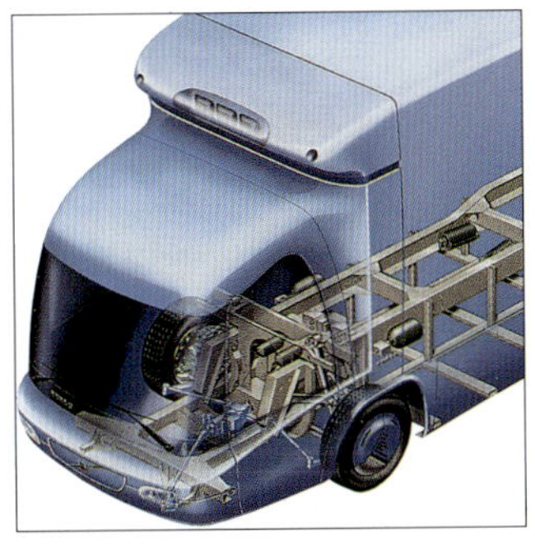

Above: *The ECT utilized an adjustable-hydraulic suspension and was made of recyclable parts.*

With increasing pressure on the global trucking industry to reduce emissions, Volvo set out in the early 1990s to develop an Environmental Concept Truck (ECT) which would serve as a test platform for new ideas.

Right: *The Volvo Environmental Concept Truck was targeted toward applications that required driving in congested urban settings. It was designed to eliminate emissions, including noise pollution.*

Specifications

Country: Sweden
Year manufactured from: 1995
Engine: gas turbine paired with electric motor and batteries.
Transmission: not available
Payload: 11.8tons (12 tonnes)
Applications: urban pickup and delivery
Special features: hybrid power reduces emissions and fuel consumption.

In addition to building a truck that would essentially eliminate emissions and reduce noise levels, Volvo engineers also set out to design a truck that would be more efficient to operate. Therefore, maximizing carrying capacity was of the utmost importance.

Designers opted for a hybrid propulsion system that would be able to utilize any type of fuel. They developed a gas turbine that was paired with an electric motor. The electric engine boasted excellent torque characteristics making the power transfer to the drive wheels very efficient.

Because Volvo was designing a truck intended for urban deliveries, the company paid special attention to its maneuverability. In the end, it incorporated an all-wheel steering system, which resulted in excellent handling.

Recyclable vehicle

Volvo designed the ECT out of parts that could be used over and over again in the construction of new trucks. As a result, the ECT was essentially a recyclable truck, and generated plenty of interest.

KOMATSU 930E
1995 USA

KOMATSU 930E

Komatsu began producing mining vehicles in Japan in 1953 when it built its first dump trucks. However, the company's claim to fame is without doubt, the 930E, developed in 1995.

This dump truck was, and still is, the largest in the world. The 930E was developed by Komatsu Dresser Company which was renamed Komatsu America International Company in 1996.

The 930E weighed in at more than one million pounds when loaded, and featured a payload capacity in the region of about 314.9 tons (320-tonnes). It was used primarily in mining applications for copper, gold, tar, sand, and coal worldwide. The 930E underwent several enhancements with the latest version being named the 930E-3. The -3 version featured an improved transmission and drive system components.

Electric-driven

The 930E was an electric-driven truck, which provided it with better power when climbing ramps and hills than a mechanically driven vehicle, the company said. The 930E had to be built of the toughest materials in order to handle weights of up to 320-tonnes (314.9 tons). The truck featured high-tensile strength and abrasion-resistant steel castings at critical stress transition zones, which allowed the truck to endure the work required of it.

Above: *The Komatsu 930E is the largest dump truck in the world.*

Specifications

Country: USA

Year manufactured from: 1995

Engine: Komatsu 2,700hp (2,013.39kw) SSDA16V160 16 cylinder engine

Transmission: not available

Payload: up to 314.9 tons (320 tonnes)

Applications: mining

Special features: behemoth mining truck weighed nearly one million pounds when loaded.

VOLVO

1996 Sweden

VOLVO VN

Volvo had been serving the North American trucking industry since the 1950s, but in 1996 the company went a step further with the introduction of its VN line of conventional highway tractors.

Right: *Volvo took a serious run at the North American heavy-duty long-distance haulage market with the VN. It featured a conventional design with a long, sloping aerodynamic hood.*

Specifications

Country: USA
Year manufactured from: 1996
Engine: full range of Volvo and Cummins engines
Transmission: full range of automated and mechanical transmissions
Payload: approx 19.6 tons (20 tonnes)
Applications: long-distance on-highway
Special features: huge sleeper provides comforts such as TV, microwave, and clothes closet.

The long hood gave the VN a distinctly North American appearance and a further benefit of the conventional styling is the flat cab floor, since no engine tunnel is required. That flat floor helped contribute to a roomy cab. When first introduced, the VN also included two sleeper cab options of different lengths, in addition to a day cab.

The following year an even larger sleeper cab was offered for the Volvo 770, which included options such as a TV set, microwave oven, clothes closet, and other features. The roomy 770 would be succeeded by the VN 780 when the line was redesigned in 2002.

Rigorous safety testing

Although the VN and 770 were designed with the North American market in mind, the Swedish truckmaker required them to pass the demanding Swedish cab safety tests. This made the truck one of the safest Class 8 vehicles to ever travel North American highways. Like most North American trucks, the VN series was available with a wide range of engines and components. The Volvo D12 engine was a popular choice, an engine that was comparable to the powerplant found in the European FH12.

Volvo wasn't content to rest on its laurels. In 2002, a further 1,500lb (680.2kg) was trimmed from the vehicle, new stylish headlights replaced the standard headlights, and the engine compartment was enhanced to house a new low-emission engine.

MERCEDES-BENZ
1996 Germany

MERCEDES-BENZ ACTROS

As the successor to the popular 23-year-old Mercedes-Benz SK, the new Actros had big shoes to fill. However, the Actros' engineers endeavored to deliver improved driver comfort and better performance.

Above: *The Actros cab was designed to stringent safety standards and could be ordered with a driver's side airbag.*

Specifications

Country: Germany
Year manufactured from: 1996
Engine: V6 or V8
Transmission: MB Telligent gearbox
Payload: 25 to 28 tonnes (24.6 to 27.5 tons)
Applications: long-distance on-highway
Special features: stability control system helps prevent rollovers.

One of the most interesting developments to be implemented in the Actros was the Telligent engine system, which was constantly provided with key information so it could make the proper injection duration and timing decisions for each cylinder. The result of this precise engine control was high power output, low fuel consumption, and lower emissions.

The Actros also featured a Telligent gearshift system mounted on the side of the seat that could simply be moved up or down to change gears. The Telligent gearbox is able to calculate the appropriate gear and automatically chooses the right one based on torque and engine speed.

The driver could override the gearbox's decision, but in most cases the driver would simply depress the clutch to accept the decision and the truck would change gears accordingly. A rocker switch allows the driver to shift up or down half a gear and if the driver tries to shift into the incorrect gear the system will give a warning.

Safety first

The Actros' cab was tested to ECE R 29 standards and a driver's side airbag was optional. The optional Telligent roll control system automatically adjusted the shock absorbers based on operating conditions and load characteristics.

Above, left: *The Mercedes-Benz Actros incorporated new technology, such as an automated gearbox that calculated the appropriate gear based on torque and engine speed. It was a popular long-distance highway truck that was easy to drive.*

WESTERN STAR

1996 Canada

WESTERN STAR 4900 SERIES

Western Star introduced its 4900 series to the market in 1996, targeting owner-operators looking for a premium truck capable of handling the most rigorous applications.

Above: *The new family of Western Star trucks was targeted toward image-conscious owner-operators.*

Specifications

Country: Canada
Year manufactured from: 1996
Engine: full range up to 625hp (466kw)
Transmission: full range of mechanical and automated transmissions
Payload: approx 20 tons (20.3 tonnes)
Applications: vocational, long-distance, on- or off-highway
Special features: highly customizable truck.

The 4900 series became an incredibly popular truck in Western Canada where it was particularly prominent in the forestry, mining, and oilfield industries. The 4900 series consists of the flagship EX, the set-back SA, and the set-forward FA. The EX featured a long, conventional hood while the SA had a more sloped, high-visibility hood.

Because of their extreme working conditions, the Western Star 4900 series had a welded steel cab, which was then hand-painted. It was then sprayed with aluminized wax for a glossy, prestige finish. Virtually all aspects of Western Star trucks are subjected to a minute inspection at the factory to ensure the trucks are built to perfection.

Lightweight sleeper debuts

Two years after the introduction of the 4900 series, Western Star developed the Star Light Sleeper, which was comprised of an ultra-lightweight polypropylene honeycomb material sandwiched between two sheets of aluminum. The result was a sleeper box that was substantially lighter than any other on the market while also providing exceptional insulating characteristics against outside temperatures and noise. The sleeper boxes are currently built by Canadian Commercial Vehicles in Kelowna, British Columbia. The same lightweight, honeycomb material is used for the floor and the rear wall on Western Star day cabs as well.

KENWORTH

1996 USA

KENWORTH T2000

Kenworth, traditionally known for its conventional styling, released its aerodynamic highway tractor, the T2000, in 1996. The T2000 was designed to strike a fine balance in pleasing both driver and owner.

Above: *Twenty years of aerodynamic retooling went into the design of the T2000. Lightweight components were also used to further improve the trucks' fuel mileage.*

Specifications

Country: USA

Year manufactured from: 1996

Engine: full range 607 to 976ci (10- to 16-liter), up to 600hp (447.4kw)

Transmission: nine or 15-speeds to bevel

Payload: approx 20 tons (20.3 tonnes)

Applications: long-distance on-highway, regional hauling

Special features: sloping hood provides improved aerodynamics, radically new look.

Above, left: *The T2000 featured a huge, 75in (1.9m) sleeper making it an instant hit with long-distance truckers in the North American market. While it was a departure from Kenworth's long, square hood, its improved fuel mileage won it fans among both fleets and owner-operators.*

The truck was unveiled at the International Trucking Show where it was presented in 112 and 120in (2.84 and 3.04m) BBC configurations with 75in (1.9m) AERODYNE sleepers. The cab was bonded and fastened using aerospace technologies that provided a smooth, durable, and weather-tight design.

Twenty years of aerodynamic research

The company said the T2000 was the culmination of 20 years of aerodynamic tooling by the company's engineers. Advanced composite materials, lightweight components, and improved aerodynamics combined to make the T2000 a more efficient vehicle to operate in over-the-road applications.

The hood, windshield, fairing, bumper, fenders, and cab extension were all designed with aerodynamics in mind. Load capacity was also increased thanks to the lightweight components. While setting a new standard for aerodynamics, Kenworth also said the T2000 raised the bar in terms of comfort, performance, and reliability and was designed with reducing life-cycle costs in mind.

Drivers were pleased with the T2000's intelligent use of space and ergonomic cab design, which contributed to a better lifestyle while on the road. A thermal core floor provided excellent insulating qualities.

DAF

1997 **Holland**

DAF 95XF

DAF introduced the new 95XF in 1997, promising to deliver lower operating costs, improved reliability, and enhanced driver comfort. The 95XF was named the International Truck of the Year in 1998.

Below: *DAF vowed the 95XF was the perfect truck for heavy and long-distance hauling. The truck's engines boasted up to 530hp (396.2kw) ensuring it could live up to its reputation as a heavy hauler.*

Right: *The DAF XF's Space and Super Space cabs provided plenty of living space for long-haul drivers. More than 35 cubic feet (1 cu m) of storage space was provided with no nook or cranny going to waste.*

Specifications

Country: Netherlands

Year manufactured from: 1997

Engine: DAF or Cummins up to 530hp (395.2kw)

Transmission: 16-speeds to bevel

Payload: 24.6 to 68.9 tons (25 to 70 tonnes)

Applications: heavy and long-distance hauling

Special features: large cabs make it ideal truck for long-distance transportation.

The 95-series had been around for 10 years prior to the release of the 95XF. However, the 95XF took the series to a new level, especially in terms of driver comfort.

Spacious cabs available

Several previously available cab sizes were dropped on the 95XF in favor of the larger Space and Super Space Cabs. Customers wanting the smaller cab sizes could opt for DAF's more modest 85-series vehicles. Both the Space Cab and Super Space Cab proved to be winners. They not only provided plenty of space, but were also very quiet.

Other features of the Space Cab and Super Space Cab included spacious lockers over the windshield, a central compartment that could accommodate a microwave oven, and storage underneath the lower bunk. There was also an airtight storage area accessible only from outside the vehicle, which could be used to store dirty or smelly objects. The total storage capacity of the Super Space Cab was more than 35 cubic feet (1 cubic metre).

The 95XF provided operators with more powerful engine choices, including Euro-2 and Euro-3 compliant engines ranging from 380 to 480hp (283.3 to 357.9kw). Despite the powerful engines and large cab, the 95XF delivered very good fuel mileage.

STERLING
1998 Canada/USA

STERLING

The Ford line of heavy-duty trucks was acquired by Freightliner LLC in 1998 and rebranded as Sterling. The Sterling brand was able to establish its own distribution network and unique product offerings.

The rejuvenated company immediately set out to become a strong provider of vocational, long-haul and owner-operator-oriented Class 5 to 8 trucks in North America. Right-hand drive Sterling trucks were also available in Australia and New Zealand. The company prided itself on the ergonomics, safety, and productivity offered within the Sterling line. All Sterling conventional trucks were produced at the company's St. Thomas, Ontario plant in Canada. About $60 million Canadian dollars was required to get the plant adapted before production could begin there.

The Acterra

The Acterra included a complete line of Class 5 to 8 trucks introduced in 1999. The Acterra's premium cab offered the same comfort, efficiency, strength, and durability as the cabs on Sterling's heaviest models. Even the medium-duty offerings exhibited the

Above: *In 1997 the Ford nameplate began disappearing off North American roads, replaced by the Sterling brand. Right-hand drive trucks were also built for markets such as Australia and New Zealand.*

traits of a heavy-duty truck rather than those of a pickup truck. For instance, the Acterra medium-duty features options such as a 120,000psi frame, 46,000lb (20,861.6kg) tandem axles and engines rated up to 350hp (260.9kw).

The Acterra was a highly customizable truck that could be tailored to meet the needs of virtually any application. Common applications for which the Acterra was well suited included dump, liquid bulk, distribution, landscaping and refuse. For applications in a city setting, the cab was designed to make it easy to hop in and out of the truck thanks to wide, self-cleaning steps and an extra wide door opening.

The Acterra could also be driven by a series of electronic diesel engines delivering good fuel efficiency, including Daimler-Chrysler's proprietary MBE 900.

The A-Line

Sterling called on drivers to help design the company's highway tractor, the A-Line. Some of their suggestions inevitably centered on driver comfort and those recommendations such as a full 11in (28cm) of seat travel were implemented in the A-Line. Other features popular with drivers included a tilt-telescopic steering column and fingertip cruise-control buttons on the steering wheel.

A variety of engines were available to be matched with Sterling's A-Line ranging from 610ci to 964ci (10 liters to 15.8 liters) and up to 600hp (447.4kw). As with most highway tractors of this era, aerodynamics were key when designing the A-Line. A set-back front axle was also adopted to improve maneuverability and to take some weight forward, allowing for higher payloads.

Specifications

Country: USA

Year manufactured from: 1998

Engine: full range up to 600 hp (447.4kw)

Transmission: full range

Payload: not available

Applications: local delivery, vocational, long-distance, on-highway

Special features: full range of Class 5-8 trucks available.

Below: *Sterling built a variety of trucks in the Class 5-8 segments. It quickly became a popular vocational truck and was widely used in the construction industry.*

SCANIA

1998 Sweden

SCANIA 4-SERIES

In 1998 Scania released its 4-series of trucks covering a range of heavy transportation requirements from local delivery to long-distance heavy haulage. The 4-series also complied with tough environmental standards.

The 4-series was divided into four different classes, making it easy for customers to choose the vehicle that best suited their individual needs. Each of the classes boasted a high level of driver comfort and safety and was easy to modify.

The Class L chassis was designed for long-distance haulage on regular roads. This particular truck could handle payloads with gross weights up to of 59 tons (60 tonnes). It could be equipped with engines of up to 530hp (395.2kw). The Class L chassis could be fitted with low, high, high-roof, or hooded day cabs or sleeper cabs. It was the ideal truck for international road haulage.

The Class D chassis featured low maintenance costs and good fuel mileage. The powertrain, chassis, and cab were reinforced, providing excellent reliability. It was best suited for short distances on normal roads and had a maximum gross weight of up to 43.3 tons (44 tonnes). Engines of up to 385hp (287kw) could be specified and low day cabs or sleeper cabs were both options.

A model for every road type

The C class was designed for demanding jobs over short distances, and was an excellent off-road vehicle thanks to its high ground clearance. The C class could handle gross weights of up to 147.6 tons (150 tonnes) and could be powered by engines up to 530hp (395.2kw). Often, the C class was put to work in the construction of heavy special transportation sectors. The sturdy powertrain, chassis, and cab made this a reliable vehicle.

Above, top: *Driver comfort was one of the top selling points for Scania.*

Above, left: *The Scania 4-series was environmentally friendly.*

Specifications

Country: Sweden

Year manufactured from: 1998

Engine: Up to 530hp (395.2kw)

Transmission: not available

Payload: GVW up to 147.6 tons (150 tonnes)

Applications: long- and short-distance on- and off-highway, construction

Special features: compliant with new environmental regulations.

FODEN

🛠 **1998 UK**

FODEN ALPHA 2000

The name Foden returned to the heavy-duty truck sector in 1983 following a hiatus after its acquisition by Paccar. When truck production was resumed, military trucks were once again a big part of its business.

Above: *With the Alpha range Foden switched from fiberglass to pressed steel cabs.*

Right: *The Alpha reestablished Foden as a player in the heavy-duty truck market. This line of trucks was used for a wide variety of tasks ranging from long-distance transportation to construction.*

Specifications

Country: UK
Year manufactured from: 1998
Engine: Cummins or Caterpillar to 450hp (335.5kw)
Transmission: 12-speeds to bevel
Payload: 19.7 to 68.8 tons (20 to 70 tonnes)
Applications: construction, long-distance, on-highway
Special features: steel cab replaced fiberglass.

However, a civilian range consisting of the medium-weight 2000 and heavy-duty 3000 and 4000 series were also introduced. These trucks were powered by a wide range of engines including the first Caterpillar engines to be offered in the UK

This line was complimented in 1996 by the Foden 4000 XL series, which featured a raised-roof cab ideal for long-distance haulage. However, it was the Alpha—introduced in 1998—that reestablished Foden as a significant player in the heavy-duty trucking industry.

Introducing the Alpha

The Alpha was the result of a redesign of the 3000 range, which consisted of a move away from GRP cabs. A Foden Alpha 2000 was also made available. The Alpha family grew to include 4x2 and 6x2 tractors, 4x2, 6x4 and 8x4 rigids, as well as an assortment of dump trucks and mixers.

The Alpha range had a steel cab much like its Paccar cousin DAF and could be ordered as a day cab or with a sleeper berth. The truck was suitable for a wide range of applications ranging from construction work to line-haul operations. They were powered by either a Cummins or Caterpillar engine of up to 450hp (335.5kw). The Foden Alpha range was a body-manufacturer's dream as it had an abnormally clean back-of-cab profile.

chapter 9

Modern Trucking

As the 21st century approached, the trucking industry was facing increased pressure to clean up its act on both sides of the Atlantic. In Europe, Euro 1 emissions standards were introduced in 1992 followed by the Euro 2 standards four years later. These new emissions standards required all heavy-duty highway trucks to stay below stringent emissions levels. Euro 3 standards came into effect in 1999.

Above: *The Freightliner Columbia combined the best of both worlds—a conventional truck with 21st-century technology.*

Left: *The Volvo VN was more aerodynamic than its predecessor and specially designed to accommodate engines with Exhaust Gas Recirculation (EGR).*

In the United States, the U.S. Environmental Protection Agency (EPA) was implementing similar standards beginning in 2002. The EPA required engine manufacturers to reduce NOx and PM emissions substantially by that year, with further reductions required in 2007 and 2010. To comply with the EPA's 2007 targets, manufacturers were forced to reduce the emissions of NOx and PM by 98 percent compared to 1987 levels.

While the goal was the same, how the manufacturers reached their targets differed between North America and Europe. In Europe, selective catalytic reduction (SCR) was widely promoted as a means by which to meet Euro 4 and 5 standards. This technology required the use of urea, which was housed in a second tank on the vehicle and sprayed into the exhaust, eliminating harmful emissions.

Most U.S. manufacturers opted instead to use exhaust gas recirculation (EGR), which redirected the exhaust fumes through a cooling system before mixing it with air entering the engine for combustion. This technology was capable of meeting the 2002 standards and with some modifications the manufacturers were confident that when coupled with a diesel particulate filter, it could also meet the 2007 standards.

FREIGHTLINER
1998 USA

FREIGHTLINER CENTURY CLASS S/T/ARGOSY

With the turn of the century fast approaching, Freightliner introduced its Century Class S/T in 1998. The new model featured a state-of-the-art aerodynamic design and an under-the-hood air-management system.

Specifications

Class S/T/Argosy
Country: USA
Year manufactured from: 1998
Engine: full range, Detroit Diesel, Mercedes-Benz, and Caterpillar
Transmission: nine or 12-speeds to bevel
Payload: 19.7 tons (approx 20 tonnes)
Applications: long-haul, on-highway
Special features: COE Argosy able to haul 58-ft (17.6-m) trailers, increasing load capacity.

Above: *With very few cab-over-engine trucks running American highways, Freightliner gambled with the introduction of the Argosy. It allowed operators to haul longer loads, but conventional-style trucks would remain dominant in North America.*

The Mercedes-Benz wind tunnel in Stuttgart, Germany, helped engineers reach new heights when it came to aerodynamics and fuel efficiency, and designers on both sides of the Atlantic contributed to the new model. Safety enhancements such as the introduction of driver's side air bags to the heavy-duty truck market were also achieved.

Aerodynamic hood

The Century Class featured a low, aerodynamic hood, which provided improved visibility and better fuel consumption as a result of improved air flow over the front of the truck. It was a lightweight, yet strong hood constructed of a sheet-molded compound. A larger windshield and mirrors also contributed to improved visibility from the driver's seat.

One of the more cutting-edge developments found inside the cab was a Driver Message Center, which provided information to the driver about such things as fuel economy, mileage, fluid levels, and other onboard systems. It also provided timely information about any attention or repairs the truck required.

MACK
1999 USA

MACK VISION

Mack introduced its premium highway tractor—the Vision—in 1999. It was designed to provide excellent handling and driver comfort as well as low operating costs and excellent efficiency.

Above: *The Mack Vision was a premium highway tractor with excellent aerodynamics.*

Specifications

Country: USA
Year manufactured from: 1999
Engine: Mack ASET
Transmission: 12 or 16-speeds to bevel
Payload: 19.7 tons (20 tonnes)
Applications: long-haul, on-highway, heavy bulk, dump
Special features: latest generation of engines feature EGR to reduce emissions (as of 2002).

The Vision featured an aerodynamically designed hood, grille, and cab, giving it a modern highway-tractor appearance. Drivers enjoyed the ergonomic layout of the cab, which included an electronic wraparound dashboard.

The Vision also provided a smooth ride thanks to the air-suspended cab and sleeper combination. The sleepers were available in a range of sizes including 48in (1.2m) flattop, 56in (1.4m) flattop, 60in (1.5m) mid-rise, and 70in (1.7m) mid-rise. The Vision was also available as a day cab for those operators who wanted a lightweight vehicle and didn't require the extra living space afforded by a sleeper cab.

Vision day cab

The Vision day cab featured an axle-back chassis and could be customized with a variety of axles, transmissions, and suspensions. One such suspension was the MaxAir, with 12,000lb (5442.1kg) front and 40,000lb (18,140.5kg) rear axles. For operators planning on putting their Vision to work in flatbed, dump trailer, and heavy-bulk applications, two frames were available including inside channel reinforcement.

RENAULT
🛠 **1999 France**

RENAULT MASCOTT

Renault first introduced its Mascott in 1999. The Mascott was a small truck that combined big-truck characteristics and hauling capabilities with passenger-car handling.

Specifications

Country: France
Year manufactured from: 1999
Engine: Renault up to 160hp (119.3kw)
Transmission: Renault
Payload: 1.97 to 2.95 tons (two to three tonnes)
Applications: regional pickup and delivery
Special features: first truck in its class to debut acceleration slip control.

Above: *The Mascott was an ideal truck for local pickup and delivery applications. It could haul loads up to 6.3 tons (6.5 tonnes) and was small enough to easily maneuver in urban areas.*

Resembling more of a van than a heavy-duty truck, the Mascott was nonetheless able to haul loads to gross up to 6.3 tons (6.5 tonnes). The Mascott was available as a van, chassis cab, chassis double cab, or chassis cowl, making it a diverse vehicle capable of many applications.

New options in 2004

In addition to making the Mascott available in a wider range of colors, Renault also improved the overall look of the truck during its 2004 redesign. A robust hood, large wraparound bumper, and beehive front grille were among the enhancements. New headlight lenses upgraded headlight performance and improved the appearance of the front of the truck. The Mascott's cab interior was also redesigned to provide increased driver comfort. For instance, the gear lever was placed at dashboard level and an arm rest was added to the driver's seat. The Mascott was also fitted with an on-board computer, which provided information such as fuel consumption and distance traveled.

Mascott was available with two new engines when redesigned in 2004, including a 160hp (119.3kw) offering, which was the most powerful engine available in that class of vehicle.

FREIGHTLINER
🛠 **1999 USA**

FREIGHTLINER COLUMBIA

Freightliner LLC sums up its Columbia truck this way: "Traditional Truck—21st Century Technology." Introduced in 1999, it features the styling of the new century with the traditional values of a Class 8 vehicle.

Above: *This was soon established as one of the best-selling highway trucks in North America.*

Specifications

Country: USA
Year manufactured from: 1999
Engine: full range of Caterpillar, Detroit Diesel and Mercedes-Benz with up to 525hp (391.4kw)
Transmission: 12-speeds to bevel
Payload: not available
Applications: long-haul, on-highway
Special features: aerodynamic styling makes it ideal fleet truck.

The Freightliner Columbia is one of the best-selling Class 8 trucks in the U.S. and Canada. Drivers like the large chrome grille, which provides excellent cooling for the engine while adding flair to the look of the truck. Advanced aerodynamics make for a sleek-looking vehicle that handles well on the highway and provides good fuel mileage. The low, sloping hood also provides the driver with good visibility, and even the fold-away mirrors are designed for optimum aerodynamics.

Seven-sleeper cab configurations are available on this North American highway truck including the 34in (0.86m) Mid Roof, 48in (1.21m) Mid Roof, 58in (1.47m) Mid Roof, 58in (1.47m) Raised Roof, 70in (1.77m) Mid Roof, 70in (1.77m) Mid Roof XT and 70in (1.77m) Raised Roof. The Columbia is also available with a day cab.

Engine options

For the Columbia engine options include offerings from Caterpillar, Detroit Diesel, and Mercedes-Benz with power ratings of up to 525hp (391.4kw). Driver comfort was addressed by the use of Freightliner's EzyRider seats that featured a wide and thick back and seat cushion and a tall backrest for improved head support.

MAN

🛠 **2000 Germany**

MAN TG-A

MAN introduced its Trucknology Generation A as a versatile range built in a modular fashion, which the company said would set the standards for optimum payload. The MAN TG-A won Truck of the Year in 2001.

The "Trucknology" phrase was coined by MAN executives to describe the advanced engineering and design that went into the TG-A. The truck was launched amid much fanfare in the year 2000 to an audience of more than 8,000 guests at simultaneous events across the United Kingdom.

Modification of the TG-A was still possible thanks to an assortment of wheelbase and overhang combinations available from MAN. Custom bodies could be easily modified to the requirements of a specific job.

Heavy payloads

The product range included a 6x2 chassis, which was capable of taking on jobs involving extremely heavy loads, and five different cabs were available so customers could choose the configuration best suited to their application and haulage distances. MAN referred to this option as a "space travel" program for both payload and driver. The MAN TG-A boasted a custom-fit drivetrain and a choice of four Euro 3 compliant engines. Six-cylinder powerplants were the norm, capable of up to 460hp (343kw) while yielding high torque outputs. Despite the power of these engines, they ran very quietly.

A strong, yet weight-optimized frame helped improve payload capacity on the TG-A. Drivers enjoyed a pneumatic spring system provided enhanced driving stability and a comfortable ride.

Above: *The MAN TG-A trucks, such as these 410 models were known for their high payloads, as they were comprised of lightweight components and materials.*

Specifications

Country: Germany
Year manufactured from: 2000
Engine: MAN 732 and 781ci (12 and 12.8 liter)engines
Transmission: 12 or 16-speeds to bevel
Payload: not available
Applications: long-haul, on-highway
Special features: optional aerodynamics package reduces fuel usage.

FREIGHTLINER

2000 USA

FREIGHTLINER CORONADO

As the 21st century began, soaring fuel and insurance costs made it more and more expensive to operate Class 8 trucks in North America and management was constantly looking for the best cost-per-mile vehicles.

Above: *Freightliner introduced the Coronado to meet the tastes of owner-operators.*

Specifications

Country: USA
Year manufactured from: 2000
Engine: full range of Detroit Diesel and Caterpillar, up to 525hp (391.5kw)
Transmission: 13-speed
Payload: 20.3 tonnes (20 tons)
Applications: long-haul, on-highway
Special features: traditional styling

Not all truck customers, however, placed so much emphasis on operating costs. For the proud breed of trucker known as the owner-operator, it was equally important to drive a vehicle that turned heads on the highway and provided a comfortable home away from home. With this in mind, Freightliner (which had been building popular fleet trucks) introduced the Coronado—an eye-catching vehicle that set out to offer the best of both worlds.

The Coronado was designed for discriminating owner-operators and it coupled traditional styling with the latest advancements in comfort, ride, productivity, up-time, and safety.

Early generation throwback

In some senses, the Coronado was a throwback to an earlier generation of truck, with its traditional long-nose hood and set-forward front axle. Underneath that long hood, the largest Class 8 engines available could be accommodated satisfying the owner-operators thirst for power. Freightliner's goal when designing the Coronado was to "Create a new truck that's true to the soul of the independent trucker."

VOLVO

2000 USA

VOLVO VHD

Volvo's North American vocational lineup, the VHD, was introduced in 2000. Constructed of the toughest components, the truck was designed to handle severe-service applications such as construction.

Specifications

Country: USA
Year manufactured from: 2000
Engine: Volvo VE-345hp (257.3kw)
Transmission: Eaton Fuller 10-speed
Payload: approx 15 tons (15.2 tonnes)
Applications: construction, refuse, mixer
Special features: high-strength construction.

The cab was constructed of high-strength steel, providing more durability and toughness than aluminum or other composite materials for an extended cab life. A heavy-duty front bumper and breakaway mirrors were also used to help extend the life of the truck and reduce downtime in the event of the inevitable bumps and scrapes that occur in vocational applications.

Smooth ride

Because the VHD would often be operated in congested construction sites or within the confines of North American cities, the truck was built with a tight turning radius for improved maneuverability. In the 21st century, customers were demanding improved ride even from construction vehicles and Volvo delivered this with its T-Ride suspension, which provided a smoother ride even when operating off-road.

Also, with a driver shortage plaguing the trucking industry in the U.S. and Canada, Volvo strove to make its vocational lineup enjoyable to drive by implementing additional home comforts in the cab. An ergonomically designed instrument panel displayed necessary gauges and controls within easy reach of the driver while a fresh-air filter kept dust out the cab.

Above: *Volvo entered the North American vocational market with its VHD line of trucks. While the truck was designed to take on the most rigorous tasks, it was also comfortable and quiet inside the cab.*

INTERNATIONAL
2001 USA

INTERNATIONAL 4000 SERIES

After seven years of development and two years of fleet testing, International Truck and Engine Corp. introduced its new 4000 medium-duty series to the North American market in 2001.

Right: *International's line of medium-duty trucks would become among the most popular trucks in their class in North America. The trucks could be fitted with a variety of bodies to suit a wide variety of applications.*

Specifications

Country: USA
Year manufactured from: 2000
Engine: International DT-series
Transmission: full range of Fuller and Allison transmissions
Payload: eight to 12 tons (8.1 to 12.2 tonnes)
Applications: pickup and delivery, utility, beverage, towing, and emergency service.
Special features: quieter cab reduced engine noise by 33 percent.

The 4300 and 4400 medium-duty trucks were touted as ideal rigs for pickup and delivery companies as well as utility, beverage, towing, and emergency service applications. One of the main selling points of the 4000 series was the lack of engine noise in the cab. International reduced noise levels by 33 percent compared to its predecessor. The interior of the cab wasn't just quieter, it was also roomier. The 4000 series offered drivers up to an additional 4.5in (11cm) of belly room, an extra 1in (2.54cm) of headroom, as well as an extra 2in (5cm) of cab width.

Further improvements

The little details were not ignored either. Improved pedal placement was designed to help reduce driver fatigue, and a rear air suspension came as standard for the same reason. In the never-ending quest for increased efficiencies, International adopted a lighter, stronger frame system that increased payload capacity.

UNIMOG
2003 Germany

NORTH AMERICAN UNIMOG

The long history of the Unimog continued into the 21st century when Freightliner LLC announced it was going to be introducing the historical vehicle to the North American market.

The new Unimog would continue to be built in Germany, with distribution by Freightliner beginning in January 2003. The main applications it was intended for in the North American market included snowplows, salt spreaders, street sweepers, and brush-cutting equipment, as well as implements used by the utilities industry including aerial man lifts, tree trimmers, augers, and cranes.

Two GVWs available

The Unimog was touted as a tough vehicle that came with four-wheel drive as standard and with axle and interaxle differential locks available, making it a vehicle capable of handling even the most difficult North American terrain. It was offered in two gross vehicle weights—26,000 and 33,000lb (11,791.3 and 14,965.9kg)—and was powered by a Mercedes-Benz MBE900 electronic six-cylinder diesel engine available in 230 or 280hp (171.5 and 208.7kw) ratings.

The standard transmission was the Mercedes-Benz eight-speed, but an optional 16-speed deep reduction transmission was also available for more extreme-duty applications. A ground clearance of 18in (45.7cm) made it ideal for customers needing a vehicle for off-road work or fighting forest fires. Incredibly, the Unimog was also able to tackle 80 percent inclines. Operators could choose an option that would enable them to move the steering wheel from one side of the cab to the other, making it an ideal winter snowplow or summer shoulder mower.

Above: *The U500 was marketed as a niche truck that could be fitted with a snowplow, brush cutter, street sweeper, or other specialized apparatus.*

Specifications

Country: Germany
Year manufactured from: 2001
Engine: Mercedes-Benz MBE900
Transmission: Mercedes-Benz 8 speed standard with 16 speed available
Payload: up to 6.9 tons (7 tonnes)
Applications: off-road service applications
Special features: steering wheel can be moved from one side of cab to the other.

MACK

2001 USA

MACK GRANITE AND FREEDOM LINE

Mack's Granite had big shoes to fill when introduced as a replacement for the 38-year-old RD. The RD was legendary in terms of toughness and longevity, but Mack introduced the Granite as an equally rugged truck.

Above: *The Granite was introduced as a workhorse that could tackle vocational applications.*

Specifications

Country: USA
Year manufactured from: 2001
Engine: Mack ASET
Transmission: Mack 8 speed
Payload: approx 14 tons (14.2 tonnes)
Applications: construction
Special features: toughest cab ever built by Mack.

Unveiled in 2001, Mack actually introduced two Granite models, the regular Granite, which was designed for regions that limited weight by axle, and the Granite Bridge Formula, which was destined for regions requiring axle weight limits and increased distances between axles. The Granite's cab was based on Mack's CH and CL cabs, but $17 million later it was enhanced for severe-service applications. It was situated on top of two cushions of air and two shock-absorbers, which contributed to a smooth ride.

While toughness wasn't sacrificed, the company made the vocational truck extremely lightweight, offering customers the opportunity to increase payload and also profits.

Freedom line introduced

Mack also introduced the Freedom line of cab-over-engine medium-duty trucks. Four Freedom series models were released: the M, L, XL, and XXL. Each was available in one of two sizes with a bumper to back of the cab (BBC) length of 63in (1.6m) or 79in (2m).

AUTOCAR
2001 USA

AUTOCAR XPEDITOR

In 2001 Grand Vehicle Works (GVW) Holdings LLC acquired the Xpeditor truck line from Volvo Trucks North America. This deal also included the historic name "Autocar."

Above: *The Xpeditor was one of the few cab-over trucks built and sold in the U.S. It was commonly used for refuse hauling and airport refueling.*

GVW took advantage of this acquisition by naming the new subsidiary Autocar LLC, and charging it with the responsibility of marketing and manufacturing the Xpeditor. The Xpeditor is a workhorse of a truck used mostly for refuse hauling.

Autocar was born more than a century ago as Pittsburgh Motor Car Company. Its first vehicle was a gasoline-engined tricycle. By 1905, the company had been rebranded as Autocar and two years later it began production of trucks, beginning with a 2-tonner truck with a shaft-driven axle. In 1908, the company produced its first commercial vehicle and it forged ahead along that path in subsequent years, establishing itself as a producer of high-quality commercial vehicles.

Achieving milestones

Along the way, Autocar was the first to achieve many accomplishments within the commercial trucking industry. It was the first manufacturer to mass produce a truck with forward control within the U.S. This refers to a configuration where more than half the vehicle's length is behind the furthest front point of the windshield base and the steering axle is in the first quarter of the truck's length. Autocar was also the first truckmaker to offer a four-cylinder engine forward-control truck. It was among the first manufacturers to offer its own engines, transmissions, and rear axles. Autocar also produced vehicles that played a crucial role in both World Wars.

In the 1950s White Motor Car Company began marketing Autocar trucks, which at this time were used primarily in the construction, logging, mining, and oil industries. In 1981, Volvo bought White and took over the Autocar nameplate which celebrated its 100th birthday in 1997.

The Xpeditor

The Xpeditor is a low cab-over-engine (LCOE) Class 8 truck commonly operated in niche applications such as refuse hauling, recycling, and airport refueling. In the U.S., the LCOE segment of the trucking industry represented only about 3 percent of Class 8 truck sales when Autocar took over production of the Xpeditor.

While most other LCOE manufacturers offered a variety of other vehicles, Autocar felt it could dominate this market by focusing solely on the often-overlooked market segment. Production of the Xpeditor was moved to Autocar's factory in Hagerstown, Indiana, early in 2004. Two versions of the Xpeditor were produced there, including the WX and WXLL.

The Xpeditor WX and WXLL

The Xpeditor was built on a sturdy frame with rails made of 110,000-psi heat-treated steel joined by back-to-back aluminum cross-members. The steel cab is fully welded to deliver superior strength and its low profile is designed to allow easy entry and exit. It is capable of handling extreme-duty applications. The WXLL is a low-entry version of the Xpeditor, while the WX features a step-up cab. Originally the Xpeditor was available with Volvo powerplants but these were later replaced with a choice of engines from Caterpillar and Cummins.

Autocar also promptly improved on the Xpeditor's interior cab design by providing an extra 3in (7.6cm) of belly room and 4.5in (11.4cm) of knee and leg room. For customers who had purchased their Xpeditor's prior to these enhancements, a kit was made available, which could be installed to bring older versions of the truck in line with their more spacious successors.

New cement-mixer truck added

Autocar broadened its Xpeditor truck lineup in 2003 when it introduced a new cement-mixer truck at the World of Concrete trade show in Las Vegas, Nevada. The truck was an Xpeditor WX LCOE, which the company claimed would deliver unmatched off-road traction, stability, and productivity. It featured a Dana-Spicer tire pressure-control system. This system allowed drivers to alter their tire pressures from inside the cab in order to handle the terrain more efficiently.

Specifications

Country: USA

Year manufactured from: 2001

Engine: full range of Caterpillar and Cummins

Transmission: full range

Payload: approx 15 tons (15.2 tonnes)

Applications: refuse, recycling, airport refueling

Special features: Autocar becomes oldest vehicle nameplate in U.S. history

Left: *The Xpeditor's low cab design made it convenient for drivers to climb in and out repeatedly throughout the day.*

DAF

2001 Holland

2001 DAF CF SERIES

DAF launched its new-generation CF series in the spring of 2001. The truck, available for a wide variety of applications, featured an interior the company compared to that of a luxury passenger car.

Above: *The DAF CF was popular among drivers thanks to its advanced wheel geometry and steering system.*

Specifications

Country: Netherlands

Year manufactured from: 2001

Engine: full range from 360ci (5.9-liter) Cummins to the more powerful 769ci (12.6-liter) DAF XE engine

Transmission: not available

Payload: not available

Applications: full range

Special features: interior designed to reflect luxury passenger car.

Above, left: *The DAF CF was capable of hauling heavy payloads thanks to its strong, yet lightweight chassis.*

Business-wise, the truck was designed to deliver excellent reliability and low running costs coupled with a high payload capacity thanks to its low intrinsic weight. Fuel consumption was also a major consideration during development and a powerful six-cylinder in-line engine was paired with the CF to ensure excellent fuel mileage. Three different engines were available for the CF series, ranging from the 360ci (5.9-liter) PACCAR engine to the more powerful 764.8ci (12.6-liter) DAF XE engine for the larger of the CF models. All engines met Euro3 emissions standards.

Flat steel chassis

The steel chassis offered superior strength without excess weight, delivering a high net load capacity. Because the CF was intended for such a wide variety of applications, the chassis was designed to be built on easily. The completely flat chassis was as multifunctional as possibly, making it an ideal base for everything from a dump truck to a concrete mixer. The CF was also available in a wide range of wheelbases and axle configurations, making it a diverse truck.

The front and rear axles were equipped with DAF's disk brakes featuring flat disks which were thermally insulated from the hub to ensure a long life. The CF was touted as a truck with superior maneuverability and drivability, thanks in large part to its firm, stable steering properties.

WESTERN STAR
2002 USA

WESTERN STAR LOWMAX

For drivers who prefer to be low to the ground, Western Star introduced the LowMax 4900 EX, designed with owner-operators in mind. The LowMax sits nearly a foot (30cm) lower than standard configurations.

Right: *Many North American drivers wanted a heavy-duty truck with a low profile. The Western Star EX 4900 LowMax fitted the bill, and boasted plenty of chrome and stainless steel.*

Specifications

Country: USA
Year manufactured from: 2002
Engine: Cat C-15, Detroit Diesel Series 60 up to 625hp (466kw)
Transmission: Eaton Fuller 13, 18 speeds
Payload: approx 20 tons (20.3 tonnes)
Applications: highway applications, grain, cattle, cars
Special features: extra-low profile for retro appearance.

The tractor features an extended hood, dual exhaust pipes, and a variety of chrome and stainless-steel accessories. These attributes make it a popular show truck at "show'n'shine" events. On the road, the LowMax is known for its good handling and stability as a result of its low center of gravity. Operators who want to get the truck as low to the ground as possible can reduce its overall height to just over 104in (2.6m) by removing the cab lights and horns and equipping the truck with low-profile tires. The low ride height is achieved largely due to the low-ride front suspension and 6.38in (16.2cm) ride height AirLiner rear suspension. In addition, the Dana Spicer E-13201 front axle has a 5in (12.7cm) drop. The LowMax is available in a 132in (3.3m) BBC dimension with a set-forward front axle.

Owners of the LowMax rave about its ease of entry and exit due to its low ride and step height, and serviceability is also simpler with key parts and components more easily reachable. Despite its long, conventional-style hood, Western Star say they have reduced the frontal area of the LowMax to deliver improved aerodynamics. The LowMax can be customized with sleeper options.

VOLVO
2002 Sweden

VOLVO VN

In 2002, Volvo Trucks revamped its popular North American highway tractor series—the VN. The driving force behind the changes was the U.S. Environmental Protection Agency's stringent new emissions standards.

Specifications

Country: USA/Sweden
Year manufactured from: 2002
Engine: Volvo D12, Cummins ISX
Transmission: 12-speeds to bevel
Payload: approx 20 tons (20.3 tonnes)
Applications: highway applications
Special features: Exhaust Gas Recirculation used to meet new U.S. emissions standards

Above: *Stringent new environmental standards and the introduction of EGR (Exhaust Gas Recirculation) engines led Volvo to carry out extensive design changes to its VN range for 2002.*

The VN had to be redesigned to accommodate an Exhaust Gas Recirculation (EGR) system. Like other North American manufacturers opting to use EGR to meet the new emissions standards, Volvo conceded there would be some reduction in fuel mileage, so enhancements to the VN's cab and chassis were implemented to minimize the effect the EGR engines would have on fuel economy.

Improved aerodynamics

The cab and fifth wheel were repositioned and new headlight and fender designs were introduced to improve aerodynamics. Even the splash guards at the front of the truck were reengineered to prevent dirt buildup, which could impede airflow. In the end, Volvo claimed it had reduced drag on the new VN lineup by 3.2 percent compared to the previous models. That could result in improved fuel mileage of up to 1.5 percent, helping recoup the fuel economy loss that was inherent with the first generation of EGR engines.

The redesigned VN was also far lighter thanks to the use of aluminum and other composite materials. In fact, it weighed nearly a ton less than when the VN was first introduced in 1996. The redesigned VN lineup was built to deliver improved safety and reliability as well.

MITSUBISHI
2002 Japan

MITSUBISHI FUSO'S NEW-GENERATION CANTER

After more than eight years, Mitsubishi Fuso's Canter underwent a major overhaul in 2002. The Canter, with a payload capacity of 2 tons (2 tonnes), was redesigned to establish a new global standard.

Above: *The Mitsubishi Fuso Canter got a facelift in 2002. The new version featured a roomier cab with improved safety benefits.*

Specifications

Country: Japan
Year manufactured from: 2002
Engine: turbocharged four cylinder
Transmission: six-speeds to bevel
Payload: two tons (two tonnes)
Applications: regional pickup and delivery
Special features: collision safety technology

The new-look Canter was designed to better resemble Mitsubishi Fuso's medium- and heavy-duty offerings. It featured an inverted trapezoid grille, forward-sloping side windows and side character lines. The headlights were also revamped and the higher-horsepower models were equipped with discharge headlights that provided better visibility at night.

Dashboard-mounted gearshift

The Canter's gearshift was located on the dashboard—the first time this had been done in a mass-produced cab-over-engine truck. Mitsubishi said the new shift-lever location made shifting easier. Also contributing to driver comfort was the fact that the parking brake remained below seat level even when engaged, making it easy to slide across the cab without impediments blocking the way. The new Canter featured a shorter instrument panel than its predecessor, also allowing for more roominess.

As was the case with most truck-product redesigns, vehicle safety was also enhanced. The new Canter debuted FUSO-RISE collision safety technology, which was designed to ensure that the vehicle properly absorbs and dissipates impact energcy while maintaining sufficient space for a passenger to survive within.

DAF

2002 **Holland**

DAF—THE NEW XF95

DAF set a new standard when it introduced its 95XF model in the early 1990s, but it was about to raise the bar even further with the introduction of the new XF95 in 2002.

Specifications

Country: Netherlands
Year manufactured from: 2002
Engine: 764.8ci (12.6-liter) DAF
Transmission: 12 or 16-speeds to bevel
Payload: 25.6 to 28.5 tons (26 to 29 tonnes)
Applications: highway haulage
Special features: DAF's own disk brakes were fitted to this truck

Above: *DAF's new XF95 was designed to lower operating costs, improve productivity, and enhance driver comfort. The styling was also changed to bring it more in line with the company's LF and CF models.*

DAF altered the styling of the truck to bring it more in line with the LF and CF series (models designed for local distribution and middle-distance hauls). Redesigned, clear headlights, a new grille, and stylish spoilers helped give the new XF a fresh look. Newly designed side skirts also gave the XF95 a more modern image.

The chassis itself was completely redeveloped to provide improved rigidity at a lower weight. This was achieved through the use of new high-grade steels. The chassis was "flattopped," meaning many components could be mounted on the inside of the chassis. This subtle change created more room for components such as extra fuel tanks or a compressor for bulk transporters.

Improved efficiencies

The XF95 was fitted with a reengineered 764.8ci (12.6-liter) DAF engine available in four output versions. As a result of enhancements to the turbo and electronic engine-management system, DAF reported the new generation XFs could deliver improved efficiencies of between 1.5 and 2 percent over its predecessor. Inside the cab, the XF95 featured an easy-to-read dashboard. A suspended accelerator pedal provided extra legroom.

SISU

🛠 **2002 Finland**

SISU 8X8

Oy Sisu Auto Ab is a Finnish truck manufacturer that combines Renault cabs with mostly Mack engines for a variety of applications. This particular truck, however, is powered by a Cummins engine.

Right: *The Sisu 8x8 is a rugged truck capable of tackling rigorous jobs in extreme conditions. The Finnish Defence Force was involved in the design of the off-road truck.*

Specifications

Country: Finland

Year manufactured from: 2002

Engine: Sisu

Transmission: 16-speeds to bevel

Payload: 15.7 tons (16 tonnes)

Applications: off-road, military, peacekeeping

Special features: well protected components for off-road applications

A recent addition to its lineup was the 2002 8x8 truck designed for excellent load handling and mobility in extreme conditions. The truck was capable of taking on a variety of off-road tasks and was developed in cooperation with the Finnish defence forces. Previous Sisu off-road trucks have been used by the United Nations in peacekeeping operations in countries such as Kosovo. The Sisu 8x8 was designed to provide exceptional maneuverability both on- and off-road as well as a high load capacity. It is an extremely diverse vehicle that is capable of many different tasks in a wide range of operating environments.

Container transportation

The transportation of containers was one of the main applications for the Sisu 8x8. The truck could be equipped with a demountable platform with a Multilift hook lift attached directly to the truck's frame. The truck could then handle container-type goods independently. Sisu also incorporated an independent hydraulic system on the truck to provide increased functionality.

The truck was also designed to be easy to operate. Because the truck has to operate over any kind of terrain, the critical components such as fuel, electrical, pneumatic, and radiator systems are located in well protected places.

FREIGHTLINER
2002 USA

FREIGHTLINER BUSINESS CLASS M2

Freightliner's next generation M2 was introduced in 2002. The M2 was designed to further increase Freightliner's share of the medium-duty market in North America.

The M2 included a full line of Class 5 to 8 trucks capable of any task required of a medium-duty truck. The BBC ranged from 100 to 112in (2.54 to 2.84m) and the chassis was designed to accommodate bodies of all types. A 2,500-square inch (1.6 sq m) windshield provided a 28 percent larger unobstructed view from behind the wheel. The low, sloping hood also improved visibility. In fact, the hood was nearly as low to the ground as some light-duty pickup trucks. Since many medium-duty applications require the operator to drive in confined spaces, Freightliner also enhanced the maneuverability of the Business Class. The company accomplished this by redesigning the steering gear to allow a wheel cut of up to 55°, depending on the combination. A set-back axle also helped improve maneuverability.

Three models released

The smallest of the three M2 models, the M2 100, was ideal for driving in confined areas as

Above, top: *The Business Class M2's interior was designed with driver comfort in mind.*

Above, left: *Freightliner's Business Class M2 was used for a wide variety of jobs, including vocational jobs and fuel haulage.*

is often necessary in pickup and delivery applications. The M2 106, meanwhile, featured many of the same benefits as its smaller cousin, but was a bigger, stronger truck able to handle more rigorous tasks.

The M2 112 was the largest of the three new trucks and it was available in gross vehicle weight ratings of up to 66,000lb (29,931.9kg). This truck (also available as a tractor) had the more powerful Mercedes-Benz 4000 engine under the hood and was available with TufTrac and AirLiner suspension systems for an improved ride over rough terrain.

Mercedes power

The Mercedes-Benz 900 was specified as the standard engine for the M2, primarily due to its reputation for delivering excellent fuel mileage and excellent performance. Mercedes' medium-duty powerplant featured a competitive power-to-weight ratio and was able to produce high torque levels at low rpms. This provided good low-end driving and hill-climbing ability without requiring frequent shifts. Mercedes' more powerful MBE 4000 was also an option for the beefier M2 trucks. A variety of Caterpillar engines was also available, with power ranging from 175 to 430hp (130.4 to 320.6kw).

Attention to detail

A closer look at the Business Class M2 revealed many other small changes that collectively made the new generation superior to its predecessor. Red reading lights in the cab (similar to those found in aircraft cockpits) allowed drivers to read maps or directions at night without losing their night vision. Freightliner also revamped the wiring to eliminate unneeded wires. Multiple electrical signals were channeled through a simplified set of wires, making diagnostics and repairs to the M2 easier. A redesigned heating, venting, and air-conditioning system was also incorporated into the M2, providing a new level of driver comfort.

Specifications

Country: USA
Year manufactured from: 2002
Engine: MBE 900, 4000, Caterpillar up to 430hp (320.6kw)
Transmission: Mercedes, Eaton Fuller, or Allison
Payload: 18 tons (18.2 tonnes) with semi-trailer
Applications: regional pickup and delivery
Special features: larger window provides 28 percent larger unobstructed view

Left: *Among the Business Class M2's top selling points were its maneuverability and visibility, making it a popular truck for urban deliveries.*

IVECO

2002 Italy

IVECO STRALIS RANGE

In January 2002, Iveco introduced its Stralis Range line of heavy-duty trucks. First came the Active Space model, followed shortly thereafter by the introduction of the Active Time and Active Day models.

The Active Space was designed with the long-distance and international hauler in mind. It has plenty of cab room and a large, comfortable sleeper compartment, while the Active Time has been marketed to drivers who haul short or medium distances and don't require a large sleeper box. It does, however, include a bunk and is ideal for short trips away from home. The Active Day model is intended for regional delivery drivers who don't sleep in their truck.

2003 International Truck of the Year

The Active Space took top honors as the International Truck of the Year in 2003. The cabs of both the Active Time and Active Day were lowered 5.9in (15cm) to make it easier for drivers to enter and exit, reducing driver fatigue. The cab suspension was also modified to reduce vibration.

The major selling point for the Stralis line was its fuel consumption. Company officials boldly claimed the Active Space delivered 4 to 5 percent better fuel mileage than the competition. This was largely due to a new aerodynamic profile. Each of the Stralis Range models was also designed to reduce maintenance costs. This was achieved through the use of new engine technology.

Above: *The Iveco Stralis range included the 350 (left) and the 430 (right) heavy-duty models.*

Specifications

Country: Italy

Year manufactured from: 2002

Engine: Iveco Cursor

Transmission: automated EuroTronic 12-speeds to bevel

Payload: 27.6 tons (28 tonnes)

Applications: local delivery to long-distance international haulage

Special features: winner of the 2003 International Truck of the Year award

VOLVO

2003 Sweden

VOLVO FH16

The ongoing quest for more horsepower prompted Volvo to release its most powerful truck yet in June 2003. The FH16 became Volvo's flagship model in Europe, and featured an all-new 976ci (16-liter) engine.

Above: *Volvo's top of the range FH16 model became the most powerful road-going truck on the European market. This Swedish seven-axle outfit grosses up to 59 tons (60 tonnes).*

Specifications

Country: Sweden
Year manufactured from: 2003
Engine: 876ci (16-liter) Volvo D16C with up to 610hp (454.9kw)
Transmission: 14-speeds to bevel
Payload: up to 44.2 tons (45 tonnes)
Applications: long-distance transportation, logging
Special features: new feature, hill Start Aid.

The FH16 was designed to handle heavy loads while maintaining high average speeds in hilly terrain. Volvo was banking on higher gross vehicle weights and extended vehicle lengths being permitted in Europe that would require more powerful engines.

In Sweden and Finland, lengths of 82ft 10in (25.25m) and weights of up to 59 tons (60 tonnes) were already permitted. However, much of Europe still capped train lengths at 61ft 6in (18.75m) while most gross combination weights were limited to 39.3 tons (40 tonnes). Volvo realized higher-horsepower engines would be required by the European trucking industry. Thus, the FH16 tractor and D16C engine were born.

The truck itself was based on its predecessor, the popular FH12. However, Volvo incorporated new styling into the latest model which clearly distinguished it from previous models.

Safety features

As always, Volvo prided itself on the safety enhancements. Safety features included Active Cruise Control, which utilizes Doppler radar to maintain a safe distance between itself and the vehicle in front. The system is capable of altering the truck's speed accordingly to ensure a safe following distance. A new Electronic Stability Program (ESP) brake stabilization system helped prevent skidding and rollovers in emergency braking situations. Also, trucks featuring Volvo's Electronic Brake System (EBS) were equipped with a new feature called Hill Start Aid, which helped drivers start rolling from a complete standstill on an uphill grade.

RENAULT
2003 France

RENAULT HIGHWAY KERAX

Renault's Highway Kerax was designed to combine the comfort of a highway truck with the toughness needed for demanding off-road duties. The Kerax was built from parts that could handle off-road applications.

Above: *A rubber or air suspension made the Kerax handle well even over bumpy construction sites.*

Specifications

Country: France
Year manufactured from: 2003
Engine: Renault 673.7ci (11.1-liter) up to 307.2kw (412hp)
Transmission: 12 or 16-speeds to bevel
Payload: 20.6 tons (21 tonnes)
Applications: construction material delivery, sand, aggregates, bricks, livestock
Special features: KEEP WARM and WARM UP features keep interior and engine temperatures warm

Above: *The Kerax may look like a highway truck, but it was tough enough to tackle construction and other off-road applications as well. In fact, it was designed with the construction market in mind.*

With the slogan "Built to build," the Kerax was built specifically for the construction environment. It was also capable of hauling sand, aggregates, bricks, and tiles as well as livestock and tankers.

Low unladen weight

The eight-wheeler had a fairly low unladen weight of 19,381.9lb (8790kg) with a payload of up to 20.6 tons (21 tonnes), depending on the body type specified. The low unladen weight could be achieved by customizing a standard day cab, 5.3 gallons (20 liters) of fuel, and with a 16ft 7in (5.065m) wheelbase. The base specification included a Renault 673.7ci (11.1-liter) engine capable of 370hp (275.9kw). A 412hp (307.2kw) engine was also available.

A rubber rear provided for a smooth ride even over bumpy construction sites. Tractors in 4x2 configurations were also available with a lightweight AIRTRONIC air suspension. A well equipped day cab was standard on the Kerax with Global and sleeper cabs available as options. The cab was specially designed to provide easy access.

RENAULT

2003 France

RENAULT HIGHWAY PREMIUM 320.26

Much like its cousin the Highway Kerax, the Highway Premium boasted a lightweight chassis giving it a low unladen weight. The Highway Premium 320.26 rigid was introduced in 2003.

Above: *While its payload was slightly lower than that of the Kerax, the Renault Highway Premium was built with driver comfort in mind and was better suited for long-distance work.*

Specifications

Country: France
Year manufactured from: 2003
Engine: Renault 11.1 liter (673.7ci)
Transmission: nine or 12-speeds to bevel
Payload: 17 tonnes (16.7 tons)
Applications: construction material delivery, brick, block, and tile, sand, gravel, and aggregates and box, livestock, and tanker
Special features: refuse trucks with filters to reduce emissions.

The main difference was the Highway Premium had a slightly lower payload—nearly 16.7 tons (17 tonnes) compared to the 20.6 tons (21 tonnes) of the Kerax. A well equipped day cab was standard on the Highway Premium, and a Global and sleeper-cab options were also available. Like other Renault trucks, the company promised the Highway Premium combined driver comfort, performance, and reliability with a low cost of ownership all within a lightweight package with impressive payloads.

Standard specifications

The Highway Premium became an immediate player in the competitive lighter-duty 6x4 European market. As with the Kerax, a Renault 673.7ci (11.1-liter) engine came as standard, but the Highway Premium's standard output was slightly lower than that of the Kerax, at 320hp (238.6kw). More powerful engines were also available.

Other standard features on the Highway Premium included: a 53-gallon (200-liter) plastic fuel tank; Renault single-reduction rear axles; a three-leaf steel rear suspension, steel wheels; and a ZF 9S-109 transmission. Drivers of the Highway Premium appreciated increased headroom inside the cab. The ongoing battle to reduce emissions was well underway in 2003 and to this end Renault offered a particulate filter on Premium 320 refuse trucks. Six filter cartridges were placed in the same space as the original muffler.

INTERNATIONAL
2004 USA

INTERNATIONAL 9000I-SERIES

The International 9000i series of highway tractors consists of popular fleet trucks such as the aerodynamic 9200i as well as more conventional-styled trucks such as the International Eagle 9900ix.

Specifications

Country: USA
Year manufactured from: 2004
Engine: wide range of Caterpillar and Cummins engines
Transmission: full range of Fuller and Meritor transmissions
Payload: approx 20 tons (20.3 tonnes) on line haul
Applications: long-distance transportation
Special features: aerodynamics make it two to five percent more fuel efficient than previous models

Above: *The International Eagle became a popular long-distance truck among North American fleets. It features a lightweight, aluminum cab and an aerodynamic hood.*

The International 9200i features an aerodynamic shape that makes it between 2 and 5 percent more fuel efficient than traditional models. It has a set-back front axle to provide improved maneuverability and also offers drivers an excellent view of the highway. The truck is lightweight thanks in part to an aluminum cab—another reason for its popularity with fleets which can take advantage of this by hauling longer trailers and heavier loads. The 9200i is also popular with drivers because it boasts an abundance of storage space and plenty of room to move.

Traditional look

What separates the International Eagle 9900ix is the Eagle trim, which can be found inside and outside the cab. The Eagle 9900ix has plenty of stainless steel on the exterior, giving it a classic look as well as a full-width Texas Style front bumper.

INTERNATIONAL
2004 USA

INTERNATIONAL 5000 EXTREME-DUTY SERIES

Whether it be logging, heavy hauling, or grueling construction work, the biggest mistake a severe-service trucker can make is using an overcustomized highway tractor.

Right: *International was a player in the vocational market with its 5000 extreme-duty models. They were best suited for construction applications or other off-highway jobs.*

Specifications

Country: USA
Year manufactured from: 2004
Engine: International HT, DT, Cummins, Caterpillar
Transmission: wide range
Payload: up to 18 tons (18.2 tonnes)
Applications: logging, heavy-hauling, construction work
Special features: engines up to 560hp (417.5Kw) for heavy-duty jobs

That's why International developed its 5000 series of extreme-duty vehicles—trucks designed to handle whatever is thrown at them.

Lightweight construction trucks

The 5500i is a lightweight construction truck with an aluminum cab and optional lightweight components including the International HT 570 engine. Because of its light weight, it offers construction truckers the opportunity to boost payload and increase productivity. Its more luxurious brother, the 5600i also features a lightweight aluminum cab but inside that cab a variety of premium interior features can be found. The 5600i features a 46in (1.16m) set-back front axle for improved maneuverability around cramped construction sites.

The beefier 5900i features a 30in (76.2cm) set-forward front axle and its square, conventional hood can house engines of up to 560hp (417.5kw). It is designed to haul the heaviest of loads.

SCANIA
2004 Sweden

SCANIA R-SERIES

In 2004 Scania introduced its R series of long-haul trucks, which met the impending Euro 4 emissions standards. The trucks were designed with quality, operating economy, up-time, and driver comfort in mind.

Specifications

Country: Sweden

Year manufactured from: 2004

Engine: full range up to 420hp (313.1kw)

Transmission: 14-speeds to bevel

Payload: not available

Applications: long-distance transportation

Special features: more aerodynamic, lightweight for better fuel mileage

About 65 percent of the R-series' components were taken from Scania's existing product line ,while other parts were enhanced. The cab featured an entirely new driver interface as well as new controls and instrumentation. Leif Ostling, president and chief executive officer of Scania announced, "Scania's focus during development of the new truck range was to create an even better environment for the driver and to provide even better earning power for the owner. We have reduced weight and improved reliability and performance, while retaining that distinct Scania feeling."

Euro 4 engine available

Fuel economy was improved 3 percent due to improved aerodynamics and more intelligent control of engine auxiliaries. About 595lb (270kg) was shaved from the model resulting in improved operating efficiency. The low-emission engines produced 30 percent less NOx and 80 percent less particulate matter than pre-Euro 4 engines and were available in power ratings of up to 420hp (313.1kw) for six cylinders. Many of the components on the R-series were interchangeable with previous Scania models throughout the world.

Above: *The more aerodynamic Scania R-series improved fuel economy by about 3 percent. The new frontal styling was intended to reflect the Scania identity.*

MAN

2004 Germany

MAN 2000 EVOLUTION SERIES

MAN's Evolution series of highway trucks encompassed a full range of vehicles from the small L2000 to the special-purpose E2000 capable of gross train weights of up to an astonishing 246 tons (250 tonnes).

Right: *The MAN Evolution could haul a variety of weights, but was mostly used for short-distance haulage. The truck was lightweight, making it economical and efficient to operate.*

Specifications

Country: Germany
Year manufactured from: 2004
Engine: four or six cylinder Euro 3 or 3 compliant
Transmission: 12 to 16-speeds to bevel
Payload: eight to 10 tonnes (7.9 to 9.8 tonnes) on rigids
Applications: short-distance haulage, light long-distance, construction, utility
Special features: highly customizable

The lightest-weight L2000 featured a compact cab with a high-roof attachment for those wanting extra headroom. It was available with either four- or six-cylinder engines which met Euro 2 and 3 standards.

Versatile M2000

The M2000 Evolution could take on a wide range of tasks, but was used mostly as a short-distance haulage truck. It could also take on light long-haul transportation or be put to work locally at construction sites or for public utility services. Its lightweight design provided increased carrying capacity and made it compatible with powerful, environmentally friendly engines. Although the frame was weight-optimized, it was also stable and body-friendly.

The M Evolution trucks were available with a choice of six cabs, making it a highly customizable machine. A double cab could seat seven people while a compact cab was ideal for local delivery applications.

RENAULT

2004 France

RENAULT MAGNUM

Renault's flagship vehicle is the popular Magnum, which can be an effective highway truck operating in most applications. The Magnum is a common sight on European highways.

What makes the Magnum stand out from other highway trucks is the fact that the cab is physically removed from the driveline components. The Multipass cab sits above, and slightly separated from what the company refers to as the "technical module," which houses the engine, transmission, and other components. The result is a distinctive style and shape that make the Magnum immediately recognizable.

Other unique styling elements incorporated in the latest generation of the Magnum include a flexible spoiler with built-in foglights and trapezoid, single-piece headlights that provide excellent visibility at night.

"A true living space"

The French manufacturer refers to the Multipass cab as a "true living space" that provides drivers with an ergonomic driving position, thermal and sound insulation against outside temperatures and noise, as well as reduced warmup time for added driver comfort. The Multipass cab is designed to reflect an actual apartment, with several distinct areas including a dining room, living room, and bedroom. A swiveling passenger seat represents an armchair, complete with footrest. The latest Magnum also includes plenty of storage space so operators can enjoy more room to move within the cab. The Magnum is a popular choice among drivers because of its many comforts.

Above: *The Magnum is Renault's flagship model in Europe.*

Inset: *The instantly recognizable Diamond badge has been Renault's symbol since 1925.*

Specifications

Country: France

Year manufactured from: 2004

Engine: E-Tech six cylinder engine

Transmission: not available

Payload: 2.7 tons (2.7 tonnes)

Applications: suitable for all highway applications

Special features: Multipass cab delivers "a true living space."

MERCEDES

2004 Germany

MERCEDES-BENZ AXOR

Mercedes-Benz introduced the new Axor in 2004 to serve the 17.7- to 25.5-ton (18- to 26-tonne) haulage market. The truck boasted low weight, increased durability, high payloads, and long intervals between servicing.

Above: *Any cargo can be hauled by the Mercedes-Benz Axor.*

Specifications

Country: Germany
Year manufactured from: 2004
Engine: 364.2–728.4ci (6- to 12-liter) six-cylinder up to 428hp (319.1kw)
Transmission: nine-speeds to bevel
Payload: 25.6 tons (26 tonnes)
Applications: short- to mid-distance hauls or local delivery
Special features: available in three cockpit formats.

It was available as a trailor or as a rigid in order to accommodate a wide variety of cargo. The truck was ideally suited for short- to middle-distance hauls or local deliveries.

The Axor was powered by an efficient six-cylinder in-line engine, which delivered an impressive amount of torque. Engine sizes ranged from 364.2 to 728.4ci (6 to 12 liters) delivering up to 428hp (319.1kw). A nine-speed direct-drive transmission with hydro-pneumatic gearshift and single-plate clutch provided for smooth and efficient shifting. It was also 66.1lb (30kg) lighter than comparable gearboxes.

Cab comforts

The redesigned Axor featured a streamlined stowage concept to provide drivers with more efficient use of space and increased storage. There were also three new cab formats (haulage, distribution, and luxury) available, which were designed especially for these applications. The haulage cab featured extra storage space, the distribution cab provided more room to move and easier entry and exit, and the luxury edition included an enhanced look and feel. The S-cab was standard on the Axor, which reduced the weight of the truck significantly.

PETERBILT

2004 USA

PETERBILT 379

The Peterbilt 379 was introduced in 1986 as the manufacturer's first truck to feature an aerodynamic design. The truck has been continuously improved on to meet the demands of the customers.

Above: *The 379 has maintained its position as one of the most popular line-haul trucks in the U.S.*

Aircraft-grade huckbolts were used to put the Pete 379 together, a technique that delivers six times the strength of traditional rivets. Peterbilt's proprietary suspension systems were customizable so the truck could meet the demands of virtually any application. To make it more customizable, Pete made the 379's wheelbase adjustable in 1-in (2.54-cm) increments and provided a choice of six different frame rails so the strength-to-weight ratio was fully adjustable.

The 379 was available in both truck and tractor configurations, as a day cab or with a full line of sleeper options. It also made a back wall kit available so operators could remove the sleeper and convert the truck into a daycab if their application changed or they wanted to sell the truck in a market that was saturated with sleeper cabs.

Advanced sleepers

The sleeper options available on the 379 included lengths from 36in (91.4cm) right up to 70in (1.7m). Pete's UltraSleeper was also available on the 379, which the company referred to as the most luxurious sleeper on the road. The sleepers were touted as being exceptionally quiet and smooth thanks to Peterbilt's Unibilt Cab Sleeper System, which the company says secures the cab and sleeper into a single structural unit minimizing pitching and softening bumps.

Specifications

Country: USA

Year manufactured from: 1986

Engine: Caterpillar and Cummins up to 600hp (447.4kw)

Transmission: Eaton or Meritor up to 18 speeds; Allison automatic; Eaton or Meritor automated

Payload: not available

Applications: long-distance transportation

Special features: popular owner-operator truck

PETERBILT
2004 USA

PETERBILT 335

In 2004 Peterbilt introduced its new medium-duty 335, a truck suited to applications such as pickup and delivery, beverage delivery, construction, and refuse disposal.

Specifications

Country: USA
Year manufactured from: 2004
Engine: Caterpillar C7 or Cummins ISC up to 315hp (234.9kw)
Transmission: Eaton Fuller 6, 9, 10, or 11 speeds; Allison automatic; Eaton Fuller UltraShift
Payload: 8 tons (8.2 tonnes)
Applications: pickup and deliver, beverage, construction, refuse, tanker, wrecker
Special features: restyled fenders give sleeker, conventional look.

Above: *For urban pickup and delivery applications, Peterbilt introduced its medium-duty Model 335. Its lightweight construction and sleek, aerodynamic appearance were a departure from the traditional Peterbilt look.*

The 335's cab was made of lightweight yet durable corrosion-resistant aluminum held together with huck fasteners that were traditionally used in aircraft construction. The frame featured huckbolted frame rails providing the 335 with what Peterbilt insisted was the strongest frame available in the industry. A one-piece aluminum roof also contributed to the 335's lightweight design and helped prevent rattling while taking on demanding jobs.

Metton hood and fenders

A unique characteristic of the 335 was its one-piece hood with integrated fenders, which were both comprised of metton, a lightweight composite material that had excellent impact resistance. The aerodynamic hood provided drivers with good visibility and contributed to a sleek appearance.

Drivers also appreciated the newly designed headlights, which featured a higher intensity beam that Peterbilt claimed provided 40 percent better down-road coverage than traditional lighting systems. The headlights were protected by a durable Lexan lens casing, which was resistant to cracks and chipping.

HINO

2004 Japan

THE NEW HINO

Hino entered the North American medium-duty market in 1984 under the name Hino Diesel Trucks (USA) and has continued to make inroads into the American and Canadian markets since then.

Specifications

Country: Japan
Year manufactured from: 2004
Engine: Hino J-series
Transmission: variety available
Payload: up to 33,000lb (14,965.9kg)
Applications: pickup and delivery, municipal
Special features: abandoned traditional cab-over design in favor of conventional style

By 2004, Hino was reaching record sales in Canada and was also becoming common on U.S. roads, particularly in urban environments as a pickup and delivery or municipal vehicle. Much of the company's success was attributed to the 2003 redesign of its lineup. Hino adopted the popular conventional-style design for its North American vehicles.

In North America, Hino had several trucks available with gross vehicle weights ranging from 14,000 to 33,000lb (6349.2 to 14,965.9kg). Its North American engines incorporated exhaust gas recirculation (EGR) to meet the Environmental Protection Agency's stringent emissions standards. Hino's four- and six-cylinder J-series engines combined EGR with high-pressure common rail fuel injection and a variable geometry turbocharger. Together, these systems provided excellent fuel efficiency and low emissions.

Galvanized steel cab

Like many trucks built in the U.S., Hino incorporated components from a wide variety of suppliers, which were attached to a standard 34in (86.3cm) frame. The drivetrain components were provided by suppliers such as Eaton, Allison, Dana, and Meritor to name but a few.

The Hino cab was built to be strong, straightforward, and dependable. It provided excellent visibility and good driver comfort.

Above: *The Japanese-built Hino medium-duty trucks are used for a wide variety of jobs around the world. In the U.S. Hino switched to a conventional design in 2004 in the hopes of making inroads into the North American marketplace.*

ERF
2004 UK

ERF ECT OLYMPIC

In 2004 ERF capped off its ECT truck lineup by introducing the spacious Olympic cab to the United Kingdom. The Olympic was designed for high-specification long-haul applications.

Right: *The ERF ECT Olympic featured a high roof, which resulted in a more spacious interior, making it a popular truck among long-distance haulers.*

Specifications

Country: UK
Year manufactured from: 2004-present
Engine: Cummins ISMe-420PS or MAN Common Rail-430PS
Transmission: 12-speeds to bevel
Payload: 26 to 28.5 tons (26 to 29 tonnes)
Applications: long-distance transportation
Special features: high roof and spacious interior.

ERF tractor operators were consulted during the design process and their input was considered by engineers. The Olympic cab could be paired with either 4x2 or 6x2 chassis and customers could choose between a 420bhp (313.1kw) Cummins powerplant or a more powerful 430bhp (320kw) MAN Common Rail engine.

The Cummins ISMe-420PS was known for its fuel efficiency and drivability, as well as for its reliability.

Spacious interior

However, the main benefit of the Olympic was its high roof and spacious interior. Driver comfort was increased over other ERF ECT trucks and everything from the high-specification seats and air conditioning was enhanced. Large storage areas came as standard and a tinted windshield reduced glare and fatigue. The Olympic also featured twin bunks and ERF said it designed the cab to be a comfortable working environment.

RENAULT
2004 France

RENAULT PUNCHER

Renault introduced its Puncher truck to the roadworks and refuse dump-truck market to compliment its Premium and Midlum range of trucks. Renault touted the Puncher as a truck that could rival all others in its class.

Above: *The Renault Puncher was often used as a refuse dump truck.*

The Puncher was typically powered by a 270hp (201.3kw) dCi 6 engine with an Allison automatic transmission. A bus-style door was standard on the passenger side allowing a crew member easy access to the cab, and an air suspension provided a smooth ride.

Because municipalities are often under pressure to demonstrate environmentally sound practices, some alternative energy engines were also available on the Puncher. In Paris, about 150 Puncher NGVs were in use, which utilized natural gas engines while an electric bi-mode Puncher combining the dCi 6 engine with an electric mode was also available.

Low floor and enhanced visibility

On the job, operators frequently have to climb in and out of a refuse dump truck or road-construction truck like the Puncher, so Renault made this simpler. The Puncher's cab floor was extremely low, just 25in (640mm), making it the lowest floor on the market. The low floor height was welcomed by operators, particularly considering these applications are among the most tiring and accident-prone in trucking.

Specifications

Country: France
Year manufactured from: 2004–present
Engine: 270hp (201.3kw) dCi 6
Transmission: Allison automatic
Payload: up to 15.7 tons (16 tonnes)
Applications: road construction, refuse dump truck
Special features: low cab floor to improve cab access

SISU

🛠 **2004 Finland**

SISU EVO2

Finnish truck manufacturer Sisu unveiled a new generation of its Evo2 truck lineup at the Metko 2004 exhibition. The new enhancements were intended to be extended to Sisu's entire product line.

Right: *The Sisu Evo2 featured new, stronger axles, which could withstand heavier loads. Scandinavian log haulers took full advantage by putting the truck to work in the bush.*

Specifications

Country: Finland

Year: 2004–present

Engine: Mack E-Tech

Transmission: 16-speeds to bevel

Payload: not available

Applications: crane, task-specific

Special features: featured completely new shaft-driven FuPro axles, developed in cooperation with Sisu Axles Oy.

The Evo2 model had a new steel look, which made it appear powerful and rugged. The front of the truck featured a new air intake vent and the cooler's gratings were made of stainless steel. The air intake vent frame was also plated with chrome, giving the truck an air of luxury. If that wasn't enough to make the truck stand out from its predecessors, designers also placed a new Evo2 model logo on the front of the truck.

The front of the truck immediately proved to be quite popular and Sisu announced it would also be using a similar style on its E18 model line. The Sisu E12 Evo2 truck was fitted with a Mack engine equipped with a new Mack E-Tech engine guidance program that was custom-designed for Finnish driving conditions. The company said the development would improve the truck's fuel consumption while also improving drivability. Meanwhile the new, larger air-intake vent was designed to improve the efficiency of the intermediate cooler.

New, more robust axles

Another new development on the truck line was the introduction of shaft-driven FuPro axles, which Sisu Axles Oy helped develop. The new axles were able to better withstand the demands placed on them by the larger and more powerful Mack engine. They were also designed to handle continuously increasing full-time drive under the pressures of maximum load.

Picture Credits

All pictures supplied by Francis Dreer/Big Block except for the following:

ERF; 5:Volvo; 6: Sterling; 8: Mary Evans Picture Library; 9:Volvo; 11:Volvo; 12: Corbis/Chris Taylor/ Cordaiy Photo Library; 14(t):TRH; 14(b): Mary Evans Picture Library; 15:TRH; 16: U.S. Federal Highway Administration; 19: Daimler-Chrysler; 20: Mary Evans Picture Library; 21(t) Peter Davies; 21(b) Stilltime/Photoshot; 22(b) Daimler-Chrysler; 23: Daimler-Chrysler; 25(t): Heritage Motor Centre/Baldwin Collection; 25(b): Detroit Public Library; 26: Corbis/Chris Taylor/Cordaiy Photo Library; 27: Alvey & Towers; 28: Stilltime/Photoshot; 29(b) SAF; 30-31: Corbis/Bettmann; 32: Stilltime/Photoshot; 34-35: Stilltime/Photoshot; 37: SAF; 40:TRH; 41: Renault; 42: Heritage Motor Centre/Baldwin Collection; 44: Bob McDaniel; 45-47: Detroit Public Library; 49: Rex W. Schrader (imaginaryimages.com); 50: Stilltime/Photoshot; 51: Peter Davies; 53: Peter Davies; 54:TRH; 55(t):Volvo; 55(b): Peter Davies; 56: Heritage Motor Centre/Baldwin Collection; 58(t):TRH; 58(b): Renault; 61(both): Heritage Motor Centre/Baldwin Collection; 62-63: Oshkosh; 64(b): Peter Davies; 65-66(both):TRH; 67: Peter Davies; 69(b): Roland O. Smith; 70: Corbis; 73-75:TRH; 77:TRH; 81: Peter Davies; 83: Roland O. Smith; 84-87:Volvo; 90: Heritage Motor Centre/Baldwin Collection; 92: Peter Davies; 94(t):Volvo; 96: ERF; 97(t): Daimler-Chrysler; 97(b):Volvo; 98-99: Peter Davies; 101: ERF; 104: Rob English; 105: Detroit Public Library; 106: Heritage Motor Centre/Baldwin Collection; 107-108(both): Daimler-Chrysler; 111:Volvo; 112: Heritage Motor Centre/Baldwin Collection; 113: Sisu; 114: Pacific NW Truck Museum, Brooks, OR, USA (pacificnwtruckmuseum.org); 117: Art-Tech/Aerospace; 118-119: Daimler-Chrysler; 120(t): Peter Jarman; 120(b): Peter Davies; 121-122:Volvo; 130-131:Volvo; 133(t): Dave Gothard (davesimages.net); 133(b): Peter Davies; 134: Heritage Motor Centre/Baldwin Collection; 143: Detroit Public Library; 144(b): Andrew Morland; 146: Iveco; 147:TRH; 151(both): Oshkosh; 158(b): Kim Loeb; 162(both): Andrew Morland: 168: Scania; 169: Peter Davies; 170(b): Daimler-Chrysler; 171: Daimler-Chrysler; 172: Peter Davies; 174(b): DAF; 179: Peter Davies; 184(both)-185:Volvo; 191: Len Rogers; 192-193: Daimler-Chrysler; 194:TRH; 196: DAF; 198: Peter Davies; 201: Heritage Motor Centre/Baldwin Collection; 202:Volvo; 203:TRH; 204(b): Peter Davies; 205(b): Peter Davies; 207(b):Volvo; 208:Volvo; 211(t):Terry Harmon; 212(b):Volvo; 213: Peter Davies; 216(both): DAF; 217:Volvo; 218: Len Rogers; 223: Daimler-Chrysler; 224(b): DAF; 230: Kim Loeb; 232(t): Terry Harmon; 232(b): Robert W. Dick; 233: Heritage Motor Centre/Baldwin Collection; 240(b):Volvo; 241: Volvo; 242: Peter Davies; 244: Foden; 252(b): DAF; 256(both): Daimler-Chrysler; 259: DAF; 260(both): Iveco; 264: ERF; 265(t): Mimo Frassineti/Rex Features; 265: Peter Davies; 268: Komatsu; 270(both): Daimler-Chrysler; 273(both): DAF; 274-275: Sterling; 276(both): Scania; 277(both): Foden; 278:Volvo; 279(t): Freightliner; 280: Freightliner; 282: Renault; 285: Freightliner; 286:Volvo; 288: Freightliner; 290-291: Autocar LLC, a GVW Holdings Company; 292(t): DAF; 294:Volvo; 295: Daimler-Chrysler; 296: DAF; 297: Sisu; 298(b): Freightliner; 300: Iveco; 301:Volvo; 302(both)-303: Renault; 307: Peter Davies; 308(both): Renault; 309: Daimler-Chrysler; 312: Len Rogers; 313: Peter Davies; 314: Renault; 315: Sisu

Index

Page numbers in italics refer to illustrations